Contents

List of Figures		vi
Acknowledgements		vii
1	Introduction	1
2	The Flexible but Entirely Serious Methodology	16
3	The Theoretical Framework for the Study	27
4	Ibiza: The Research Context	40
5	Goin' Ibiza: Home Lives and the Holiday Hype	52
6	Constructing Ibiza: The Holiday Career and Status Stratification	77
7	'You Can Be Who You Want to Be, Do What You Want to Do': Identity and Unfreedom	97
8	The Political Economy: Consumerism and the Commodification of Everything	123
9	*Capitalismo Extremo*: Risk-Taking and Deviance in Context	159
10	Going Home ... Only to Come Back Out	188
11	Discussion and Conclusion	207
12	Meanwhile across the Mediterranean ... (or So Some Wish)	230
Notes		239
Glossary		241
References		243
Index		249

List of Figures

2.1	Mapping of the West End 'drinking strip' in San Antonio, Ibiza	24
8.1	Inside a strip club on the West End	151
8.2	Mapping of Blue Marlin	157

Acknowledgements

Numerous people have helped make this work possible. My first thanks must go to the Southside Crew, who let me invade their holiday and let me spend time with them. They also permitted me into the intimacies of their home life and allowed me to go out with them on Southside. In particular, Nathan, to whom this book is dedicated, was always happy to help me clarify my findings. Thanks to my publishers at Palgrave Macmillan, in particular, Julia Willan and Harriet Barker. I am eternally grateful for the assistance from Danielle Kelly, Dimitar Panchev and Dorina Dobre: I hope you learned from the experience and take this forward in your respective careers. JoyAnn Bannon also did a sterling job with transcriptions (again). Thanks to Tim Turner, Professor Keith Hayward, Lauren Holdup, Ruth White, Paul Brindley, Tara De Courcey and Kerri David for their assistance on one of the fieldwork excursions. My thanks extend to my new friend Dr Sébastien Tutenges, who also helped with this project in more ways than one: undertaking fieldwork, collaborating on papers, writing proposals and reading my draft manuscript. I am also grateful for the use of some of his field notes. In a similar vein, Professor Simon Winlow and Emeritus Professor Geoffrey Pearson also made time from their busy schedules to comment on a draft of this work. Geoff, in particular, and who has championed my work for some time, I will miss as he has sadly passed on since helping me. I have made new friends in Ibiza, in particular, across the council and police. A special thanks to Montse Juan, Amador Calafat, and Belén Alvite. I am also grateful to the radio and television networks who helped me broadcast the work. Thanks always to my colleagues at the University of East London who continue to make it a pleasure to work with them: Dr James Windle, Dr John Morrison, Barry Heard, Jennifer Petersen-Schmidt, John Strawson, Paul Kiff and Paul Stott. Thanks also to the Dean, Fiona Fairweather, for taking the gamble on the research.

Lastly, but by no means least, there are the obligatory thanks to those closest to me. I am certainly grateful for the way in which my parents did not force upon me any pressure of what I had to be in life and instilled in me a curious freedom, independence and a sense of exploration. I apologise to my wife for almost forcing upon her my requests to go to strip clubs and brothels but, at the same time, it was certainly not for the entertainment. I never realised I would be doing this sort of thing with my life and, in some ways, still cannot quite believe it. I am just grateful I have this opportunity and to have been equipped like this to undertake research and show others how the world works.

1
Introduction

> *Nathan*: It's like, this is it! [Looking around and pointing at the beach and bikini surroundings] This time won't come again so we have to take it, do it while we can. [Looking at me in wonder] We are here, mate, we are living the dream!

Introduction

This book is about the deviant and risky behaviours of a cohort of young, working-class British tourists in a resort called San Antonio on the Spanish island of Ibiza. While it is about these particular behaviours, it also addresses the wider cultural framework which guides aspects of their day-to-day lives, attitudes and lifestyles. When I say 'deviant' and 'risky', I refer to behaviours which encompass forms of substance use, sex, violence, injury and other unconventional acts which can occur on holiday. Collectively, these behaviours have implications for them as individuals but also for other tourists, local Spanish residents/businesses, the holiday resort (local economy, law enforcement, healthcare) and their consular services and home governments. It is likely that some of what I have to say about these behaviours will resonate with what goes on in other tourist resorts, but here I focus in detail on the resort of San Antonio.

The idea for this project stemmed from reading a newspaper article in May 2010 which bemoaned another summer of problems caused by British tourists abroad. After commencing a thorough search of media articles and reports, almost everything I found seemed to pathologise their behaviours as though they were the result of deeply flawed decision-making processes, entirely disconnected from other subjective, social, cultural and structural indicators which might influence their conduct. When I started to look into the academic literature on the topic, I found very little ethnographic work which sought a lived-experience perspective and insufficient detail on the political economy – or the macro socio-structural forces – and how they may help cultivate these behaviours. Much of the existing material also relied heavily on epidemiological surveys and dated subcultural analyses; the former lacking subjective depth and the latter less consistent given the blurring of class distinctions as a result of access to various economic and cultural resources, and changes to British culture and society more generally. Collectively, I didn't feel these depictions sufficiently explained how and why the behaviours were occurring and felt this gave me sufficient impetus to begin my own investigations using ethnographic methods such as participant observation, focus groups, and open-ended interviewing.

There is something distinctly liberating about the holiday as a social occasion. People say they can 'be who they want to be' as they free themselves from the shackles of home routines, responsibilities and identities: it is a transformation from the ordinary and mundane home to the extraordinary and the hyperreal pleasures on offer in the holiday resort. The resort is therefore a place and the holiday is therefore the occasion – the space and time respectively – where fantasies can be played out; a temporal moment in life in which all the 'experiences' must be seized before the inevitable return to 'normal life' beckons. Yet it is not just simply that the people in this book go abroad and drink, take drugs and go crazy because we are talking about more complex social and structural processes, and my principal arguments in this text are that, for this group of working-class British tourists, these are:

- How their norms, values and attitudes to excessive consumption have come to be shaped over time in the UK and marketed by consumer capitalism only for them to reproduce exaggerated versions of those behaviours abroad (Structuro-culturo);
- In the resort of San Antonio, they are with others doing similar/same things which reinforces that what they do is expected of them (Social);
- When away, restrictions on behaviour are loosened as deviance and risk are endorsed in a landscape designed for their excessive consumption to take place (Spatial);
- And that, related to these transitions, how identity becomes increasingly pliable to seek as much pleasure and self-indulgence in the time available (Subjective-situational).

In this book, I want to argue that these dynamics all play a part in the deviance and risk-taking of my cohort. The first thing I want to suggest is that although this group may think they are liberated from usual behavioural protocols when abroad, really what they are doing is playing out an extension – albeit an exaggerated one – of what they do at home at weekends (drinking, drug-taking, shopping, etc.). The reasons for this, I want to show, are related to what has happened over the last 30 years as we have moved into a neoliberal, free market society based around consumption. Indeed, the social consequences of this transition are visible in most town centres across the UK because attitudes to consumption are excessive as young Brits loosen up for hedonistic weekends (see Winlow and Hall, 2006; 2009). So to some extent, the *habitus* of the people in my book – a set of dispositions such as cultural tastes and life attitudes which they deploy and initiate without conscious thought (Bourdieu, 1984) – are already well secured around weekends on the town, drinking and drug-taking at raves/clubs. As Webb et al. (2002: 36–7) note, it is this predisposition which is *'gained from our cultural history that generally stays with us across contexts and allows us to respond to cultural rules and contexts in a variety of ways'*. People say they feel 'free' and can do what they want to do in Ibiza but I am asking what 'liberation' and 'freedom' comes from going to a place only to reproduce what is done at home (drink, take drugs, eat burgers, shop)? This is why in this book I am instead arguing that it is actually *unfreedom* (see Žižek, 2002). When I use this term, I mean to describe how the social system prescribes the 'freedom' to choose which club to go to, where to shop and how to fill our lives with consumerables and, in doing so, distracts us from the real issue of our freedom. So the behaviours these young working-class Brits exhibit abroad, to some degree, have been already structurally conditioned, socially constructed, packaged, repackaged and marketed to them – and it is this commercial pressure which is aggressively foisted on them during their holiday in the resort.

Through a process of socialisation, the young British tourists in this book have also come to learn what behaviours are expected of them on holiday. Firstly, they absorb this through news media, popular culture and films which collectively promote the leisure life and holiday destinations where all the celebrities are engaging in deviance, risk-taking and all-out excess. Secondly, many in this sample have also already learned that excessive consumption and self-indulging on holiday is what they should be doing, having visited other resorts around the Mediterranean (such as Ayia Napa, Faliraki, Malia, etc.). Some have therefore developed what I call a *holiday career*. In recent years, however, global corporations, commercial entrepreneurs, the Superclubs, and the music industry have ratcheted up the marketisation of Ibiza which has made it increasingly 'mainstream'. This has meant that a younger, impressionable group have taken the short cut to Ibiza's shores with little idea of what is expected of them and this makes them more vulnerable to persuasion on holiday: after all, they are also ready to party and do 'crazy things'.

Either way, when both groups arrive in the resort, they are among other people who are behaving the same/similar sort of way so the general social context acts to reinforce the attitudes and behaviours expected of them. However, once most have visited Ibiza, they learn there are 'better places' which with their attendance bring them more social kudos. They could be that 'special person' if they saved up all year round and came back next year or maxed out credit cards just to get a piece of it. But in Ibiza you can become a part of the elite as long as you are willing to part with your money. Here capital buys a very temporary crown to wear and throne on which to sit but it all means very little unless other people can know about it. This is because these days social status is affirmed through the creation of social distinction (Bourdieu, 1984); a process that is driven forward by the institutionalised envy that is such an important feature of contemporary consumer culture (see Hall and Winlow, 2005a; Hall et al., 2008). And this is precisely why some come back because what we also are seeing in San Antonio, and across Ibiza, is the consistent reinvention of space: not only to accommodate the cultural practices of these working-class Brits but also to commodify them, and, in doing so, thereby attaching to them new levels of ideological social status.

Perhaps unsurprisingly, the resort seems to offer unlimited hedonism and fun, yet there is next to no regulation on what these young Brits do: anything seems possible. Of course, the question then arises, if everything is possible, why do we consistently see the same kinds of commodified excess? Why does a supposedly boundless terrain of cultural 'freedom' inevitably devolve into excessive drinking and beating the well-trodden track to the island's expensive Superclubs? As the book unfolds, I will try to answer these questions by connecting the immediate experiences of my respondents to the background issues of cultural formation and capital accumulation. This is important because Ibiza is now in competition with other European destinations which can now easily book the same DJs and host the same sort of mega-events. This means what is permissible on behaviour needs to be ambiguous to ensure that money is spent (Calafat et al., 2010). In addition, to make sure that Ibiza's falling tourism numbers spend as much as possible in the six months in which the island is open for business, new and varied consumption spaces which permit and endorse unlimited spending and excessive consumption need to offer the services that keep the crowds returning. I will show how these spaces inadvertently function and, in doing so, promote deviance and risk-taking.

We also cannot ignore the subjective desires attached to deviance and risk-taking which underpin postmodern life (see Hayward, 2004; Ferrell et al., 2008); after all, many of the people in this book feel home life has become quite boring, quite samey. Detached from the past fault lines of community, work and family, the same group these days are individualised; charged with crafting out a 'successful' and coherent life narrative which will retain life meaning or 'ontological security'. In times of risk, insecure work futures and fractured life narratives, mainstream consumer lifestyles now often include extreme drinking and recreational drug use. These things, it seems, have become part of a broader consumer cornucopia also built around shopping and the incessant search for pleasure. For the people I have spoken to in Ibiza, full commitment to a consumer lifestyle appears to signal genuine *being-in-the-world*; that their lives count for something. To live a decent life these days means to have experienced forms of pleasure and excitement, and our desire to inject these things into our lives provides the market with the energy it needs to continue onwards. The traditional concern with living a happy and fulfilling life, in which happiness is provided by those things that endure, has been replaced by a mere concern with fleeting, and often consumerised pleasures (Hall et al., 2008). Culture now commands us to commit to solipsistic pleasures – even if it becomes painful. If we fail to do so, we appear to be condemned to a shadowy life of 'just getting by' (ibid.). It is hardly surprising that when people have time outside of the humdrum world of work and the daily grind that they place a clear emphasis on excess. Sex and drunkenness, and the adoption of forms of behaviour that are usually external to that normal run of things, come to the fore as consumers attempt to increase their experience of hedonistic abandon before they head home and return to their work in the postmodern factories of Britain's deadening and exploitative service economy.

A holiday in Ibiza is appealing because of the way these young Brits have come to believe the weather, the music and the resort offer a complete 'holiday experience'. However, once memories (or some version of them) start to be associated, stored and reflexively revisited, so then Ibiza is constructed and reconstructed as a place of 'dreams' and 'nostalgia'. And where does one go to revisit moments in which their most prized life's stories were created? The same place year after year, and it is this ideology which connects perfectly with the *habitus* of the bulk of my sample who seek to make the most of their precious leisure time: to seize it. This is also made possible by the in-between commodified and commercial means of Ibiza 'old skool' compilations, reunion parties and the branding of the clubs in the UK which unite it with Ibiza. A visit to the island has become a point of reference in real time ('I'm goin' Ibiza/I work in Ibiza') and then reflexively positioned as one of 'life's achievements'; that one has lived a truly fulfilling life or one has a story to tell when back in the boring tedium of work or to their little ones as the grey hairs sprout or the others fall out. And this is why when the 'good life' is lived in this short period, it is reflexively revisited as a micro-experience of pleasure and/or extreme experience which is accessed through memory, and this is often how it is retained on the return home. A continual pursuit of these pleasures through memory makes home life look incessantly overcast and oppressive; happiness seems far away. This may tempt them further into either returning to Ibiza on holiday or, even more permanently, to work and 'live the dream' by becoming a casual worker in the bars and/or clubs. This is the power of ideology which keeps many of these people returning year on year.

In the end, the determination to have a 'good time' and to have it put before them, often results in deviant and risky, if not fatal, consequences. Of course this is not the case for all Brits who holiday in Ibiza; not all go away to get brain-dead drunk and many do visit and return without problems. However, over the course of researching this work, a significant number of people have suffered a number of life-changing injuries, experienced sexual harassment, made accusations of rape/spiking drinks, and sadly, died out there. It has certainly led me to question the 'fun' which people say they have on the island. Indeed, when I have challenged some people on the 'appeal' and 'fun' of Ibiza, some have struggled to articulate it; instead appearing confused about why they chose to go or what was enjoyable about it even after some said it was 'great' or they had the 'best times' – but, once again, this is the ideology at play.

So the deviant and risky behaviours for which they are blamed, I want to highlight, therefore don't really evolve from an individual pathology or some lone intentions to get wasted/high/have sex with as many people as possible/beat the crap out of anyone: there are other important dimensions to consider and in this book I would like to draw these into the equation. What you are about to read may either surprise/disgust you in a number of ways, or alternatively the behaviours I discuss may be quite familiar to you; especially given that the gendered foreign holiday has become a veritable 'rite of passage' for large sections of Britain's youth population. Either way, I want to bring you closer to the experience of my participants, and because of this, there are some graphic references to sex, violence, drug-taking and other forms of deviance and hedonistic excess. The way in which I went about this study will undoubtedly cause some upset among some established social scientists because to get some of my data I drank alcohol with my participants, went to strip clubs, brothels, Superclubs, and did most of the things associated with being in a holiday leisure zone (apart from taking drugs, having sex and engaging in the casual sexism that is such a feature of commercial space in Ibiza). At times, I had to do things, which may appear to the reader as if I too was enjoying them, but engaging in these activities was simply part of 'playing the role' in this social context; one cannot simply seek to undertake a participant observation study in a holiday resort without adapting to some degree to the socio-cultural context of the tourists' norms and the expectations which come with the social terrain. For now, please meet these young men from Southside, UK who make up the central narrative of this book and who invited me into their world to experience their holiday during the summer of 2011.

Meet the Southside Crew on a typical afternoon in San Antonio

It is 3 p.m. on a June day in San Antonio. I walk along the bay where sit and sprawl hundreds of young British tourists. I am looking for groups of men to interview and approach a group of four guys sitting in a beach bar. I sit with them, and the music pumps out and reverberates through our plastic chairs as the sun shines down on our half-naked bodies while they cheer and sip cold pints of beer. They seem to be the centre of attention in the bar as a young family leave looking aggravated at their noise. As we commence the interview, they struggle to remember how much they have drunk on the plane, in which bars they have been since arriving and even where they are staying. After some debate, they conclude they had arrived in Ibiza at 10.30 a.m. that day but had been drinking without a break from 6.30 a.m. They had spent less than ten minutes depositing their bags before heading out on their bar crawl up the San Antonio Bay towards the West End 'drinking strip' ('West End' hereafter).

It transpires that this group of friends in their mid-twenties whom I call the Southside Crew – Jay, Paulie, Marky and Nathan – come from a coastal town in the UK. Another, Simon, who was due to fly out with them, was banned from the flight because he joked with security staff that he had a bomb in his hand luggage. They say he will fly out the following day. The Southside Crew have known each other since school and three of the four now work together on various temporary construction contracts while one is unemployed. They relay their experiences in the clubs back home in their teens and early twenties when they used to be *'proper on it* [drugs]*'* most weekends; *'pills, coke, the lot'*. However, since settling down with families and/or partners, they say the opportunity to go out doesn't present itself as much. At home, they confess to getting into fights and all four have been in trouble with the law: Paulie has served two years in prison for cocaine dealing and Marky three years for GBH. On their return to the UK, Jay and Nathan, who have been arrested before, also face a court case for battering some men outside a nightclub. Last year, two of the four went to Magaluf, Spain together. They tell me they have come to Ibiza for a 'blowout', and because of the 'name' and the Superclubs – where they can't really afford to go so end up most nights on the West End.

They concede to needing vodka red bulls to pick them up between moments where their drinking has started to flag since arriving. As another pint arrives, they argue with the waiter, saying that they are owed a 'free shot', and eventually seem to bully him into bringing them over. They then recount tales of sex from Magaluf, high five and hug me before we move on to talk about girlfriends:

Nathan: It's a holiday mate. I love my girl, I love my baby.
Paulie: Clearly not.
Nathan: But I am away, I need to get fucking something. If I don't have sex on this holiday I am going to go back more frustrated, more angry.
Marky: And he'll end up taking it out on the missus [then laughs].
Dan: [To Paulie] What do you think of this?
Paulie: Terrible.
Nathan: It's best to be honest mate.
Paulie: Seriously, your girlfriend is pregnant.
Nathan: But you don't get a fuck when your missus is pregnant.
Paulie: That's why I think it's terrible ... [A young woman in a thong bikini walks past] Cor, look at the tits on that!

The irony of Paulie moralising Nathan for his attitude towards women becomes apparent the moment the semi-naked woman walks past. Despite their home relationships, two concede that there is a little competition to chalk up as *'many shags as possible'*. Some more young girls approach us with make-up and fake tan melting down their faces and legs. They try to sell us club tickets but perhaps because they don't take the boys' fancy, there is little interest. After Marky recounts a tale of being ambushed by a *'black bird'* for a €5 *'blow job'* in Magaluf, we get on to the subject of bodies and they all concede to strict diets prior to the holiday. Paulie, in particular, concedes to using steroids to maintain his figure and this, he says, is why the girls approach him. This is perhaps confirmed for him as we are once again approached by PR[1] girls, vying for us to part with our money to attend their bar party. The Southside Crew all play down drug use on holiday. Despite this, it emerges that Jay just bought weed from an African man; he says *'it don't count.'* When the bill finally arrives, they argue and claim they have been ripped off but the anger quickly diffuses when their mouths drop open as half-naked girls walk past in bikinis. Jay says *'fuck it'* and throws a bunch of euros on the table.

This was my first encounter with the Southside Crew and, over the next five days, I drank with them, spent time with them in their hotel and on the beach to follow their holiday activities. Since arriving home that summer in 2011, I also kept in contact with them through mobile phone texts, Facebook messages, meetings and nights out in their hometown, Southside. They represent typical constructions of working-class British tourists abroad in resorts like San Antonio – young people, drinking and being loud, taking drugs, confessing to intentions for a 'blowout' and perhaps also by creating a headache for the bar staff (Briggs et al., 2011a). For the remainder of this short, introductory chapter, I would like to set the study in context by considering the increased prevalence of tourism across the European Union (EU) and the current literature on British tourists: where they go on holiday, the deviant and risky behaviours for which they are attributed and the extent to which they do it.[2]

Tourism across the EU

Since the 1960s, a growing number of young Europeans have chosen to spend part of their holidays partying at international holiday resorts. There are several reasons for this which include the advent of cheap air travel; the emergence of package tour operators that target young people (Hesse et al., 2008); the development of tourist destinations that promote and capitalise on nightlife activities (Briggs et al., 2011a); the increased importance of leisure and pursuit of 'time out' among young people (Measham, 2004); and increased discretionary spending power available to youth in some European countries (O'Reilly, 2000). Today, millions of young Europeans holiday abroad each year with a primary aim of partying, often in southern European destinations. The nightlife and tourism industry have capitalised on this trend, and this is evident in the way these destinations are marketed to young people but also how they are designed to accommodate their leisure pursuit of 'time out'.

Unsurprisingly, therefore, tourism plays a major economic role in Europe, generating over 5 per cent of the EU's gross domestic product and providing around ten million jobs (European Commission, 2010); hence a healthy tourism industry can bring major benefits to local and national economies. However, the kind of economic 'benefits' increasingly include vacuous hyperreal spaces which have next to no resemblance to the rich history attached to some tourist destinations and this has come at the expense of the dissolution of local tradition, culture and community life. This is particularly the case in the tourist resorts of places like Ibiza where between the barrage of hotels sit KFC, Pizza Hut and all the global chains. As we will see, the way in which these chains and brands have settled these resorts has dislocated, if not dismantled, local tradition and culture (Chapter 4). Instead, these kinds of resort tend to have large nightlife scenes which foster high levels of deviant and risk behaviours which tend to cause harm to young people. These harms include elevated alcohol and drug use, unprotected sex (with other tourists, local populations and workers in the sex industry), anti-social behaviour, crime, violence, and unintentional injury, including road traffic injuries. One study calculated that in one year across all ages, injuries sustained by non-domestic tourists in EU countries accounted for approximately 3,800 deaths, 83,000 hospital admissions, and 280,000 emergency department treatments (Bauer et al., 2005).

Deviant and risk behaviours therefore have a major impact on young people's health and wellbeing in both the short and long term, and these effects are often amplified when they occur in foreign countries where culture, language, geography, legislation and service provision are unfamiliar (Bellis et al., 2003). Moreover, the costs of hedonistic holiday behaviours do not only affect young tourists. In these resorts, inadequate policing and limited health resources struggle to manage tourist behaviours and deal with the consequences (Tutenges, 2009). Further, tourism industries can be damaged by bad press when these problems occur, leading to image and future marketing dilemmas. Local communities are also further affected through exposure to cultures of intoxication, cheap alcohol and illicit drug markets. The costs are also swallowed by holidaymakers' home countries; for example, by providing consular and diplomatic services, and treat ongoing health problems brought home by tourists (e.g. STIs – see Hawkes et al., 1997; Hughes et al., 2009).

All this has been occurring at a time when young people are becoming mobile throughout the EU, largely because of the expansion of low-cost airlines. Consequently, EU touristic destinations have had to develop to accommodate larger numbers of visitors or, if they are somewhat established like Ibiza, seek to diversify their marketing tactics to appeal to 'new' tourist populations or get the most out of those who visit. There is, therefore, an increasing need to ensure that destinations are managed safely and the infrastructures of the local economies are organised in a way which minimises potential tourist problems. Unfortunately, however, a significant number of these problems are directly made attributable to British tourists, and, for a number of years, the extent to which they occur has not diminished.

British tourists abroad: The facts

Like their European counterparts, young British tourists have benefited from cheap international airfares. Popular resorts where the British holiday can be found across the south coast of Spain and its Balearic Islands (primarily San Antonio on Ibiza, Magaluf on Majorca), Greece (Malia on Crete, Kavos on Corfu, Faliraki on Rhodes, and Laganas on Zante) and Cyprus (Ayia Napa), with emerging destinations in Bulgaria (Sunny Beach), Slovenia (Izola), and Turkey (Bodrum, Gümbet, Marmaris). While southern Spain and the Balearic Islands (Majorca, Ibiza and Menorca) have been popular for decades, in the 1970s, 80s and 90s, new tourist destinations evolved and were marketed at the British tourist. Known as 'package holidays', they started to become popular and helped establish new transport and tourism economies, while at the same time, created unanticipated competition for already-established destinations such as those in the Balearics. However, over the last 20 years, the British tourists who holiday there have developed a reputation for social problems and negative behaviours (Calafat et al., 2010; IREFREA, 2010). For example, in Spain from 2010/11 to 2011/12, assistance for British citizens rose from 4,971 to 5,406 cases with increases in total arrests/detention (from 1,745 to 1,909), hospitalisations (from

1,024 to 1,105) and total deaths (from 1,639 to 1,755). Ibiza and Majorca are continually referenced for increases in emergency admittance to hospital from road accidents, balcony incidents and heart attacks. Here I would like to provide some brief overview of the extent of the behaviours to which they are attributed in Ibiza.

The extent of British deviant and risk behaviours in Ibiza

The most recent data on the level of deviant and risk behaviours among British tourists comes from researchers from Liverpool John Moore University. Their survey of 1,022 British tourists aged 16–35 in Balearic island airports of Majorca and Ibiza found that over half experienced 'drunkenness' five days a week or more (52 per cent), a similar percentage (54 per cent) reported using drugs and around a fifth having unprotected sex (20 per cent) with multiple partners (15 per cent) while on holiday. One-third of the visitors to Ibiza were current ecstasy users (34 per cent) and cocaine users (34 per cent). Indeed, while just 2.4 per cent of ecstasy users in the Ibiza sample reported using the drug two or more times per week at home, during the holiday nine out of ten users reported use at this frequency, with 46.7 per cent reporting use five or more days per week. Indeed, in comparison to German and Spanish tourists, the British use more illegal substances in Ibiza when they holiday than when they are at home (IREFREA, 2010). The consequences of these levels of drug use are perhaps most poignantly highlighted in hospital admission statistics. For example, during the 2005 tourist season, 135 serious drug poisonings were treated at Ibiza's Can Misses hospital in August alone; 80 per cent were non-Spanish males in their 20s, and most were brought in from the island's Superclubs.

More qualitative analyses have since followed and here is where some of my work began. In 2010, I led a team of four researchers to investigate binge drinking in San Antonio, Ibiza (Briggs et al., 2011a). During that excursion, we found that excessive alcohol consumption was socially embedded as part of the holiday ambitions of various single-sex groups of young British tourists but that also this was endorsed, and at times aggressively coerced, by players in the social context in various marketing material, the tour reps, casual workers and the bar workers/owners (also see Briggs and Turner, 2011; Turner and Briggs, 2011). The following year, in 2011 when I met the Southside Crew, I documented the role of the other players of the social scenery – bar and PR workers, the strippers, lapdancers, prostitutes as well as augmented my data on the British tourists (Briggs et al., 2011b). We found that deviance and risk also emerged in the context of risky sexual practices, drug and alcohol consumption as a consequence of the 'marketisation' of sex, drugs and alcohol which was prevalent as much in the ambitions of the British tourists as it was in the discourses of the casual workers, the PRs, the club promoters and the general landscape of San Antonio.³

Indeed, the appeal of the holiday, as a continuous leisure venture, has also been recently documented through the increased number of British casual workers/tour reps. While there is some ambiguity attached to whether they are 'tourists', from my observations of what they are doing, I can only conclude that it is some extension of the holiday because often the workers do not keep the same job with the same company/boss for long, it is not permanent work nor intended as a long-term career. O'Reilly (2000: 113) would call these people *'residential tourists'* whereby *'worklessness is celebrated and the work/leisure distinction is blurred'*. And this is perhaps reflected in quantitative studies which have found that such workers are more likely to be using more drugs, and consider it safer to do so, than the British tourists (Hughes et al., 2009). Like the tourists, the casual workers find it difficult to fend off the temptation to party – probably because they have been told it is part of their job to stimulate the celebration and encourage drinking, sexual-innuendo-like games and general deviance (Guerrier and Adib, 2003). Recent research in Ibiza has found that this particular band of Brits are more likely to take more drugs, take more risks with unprotected sex with multiple partners and have sought healthcare services abroad than the tourists (Kelly, 2011; Briggs et al., 2012b). It is these exposés which trigger both local authority responses in Ibiza and national governmental initiatives at home: both of which aim to 'responsibilise' the tourist in the hope that they might regulate their own behaviour rather than challenge problematic strains in the political economy.

Rationale for this text

I am writing this book because the way in which young, working-class British tourists are portrayed in media and academic research lacks appropriate structural and social context; in general the behaviours appear as pathological faults of the individuals who engage in them and/or as part of some sort of dated subcultural reaction to their social position. In particular, there has been a lack of detailed, ethnographic enquiry into the context of these behaviours and attributions made towards the reasons 'why'. I am also cautious of the way in which some liberal commentators would suggest that those who holiday in places like Ibiza are making 'free will' decisions and continue to make such clear-cut decisions to regulate their behaviour when they are there. I am therefore interested in removing the ideological blanket that covers the norms and value systems of my participants while, at the same time, exposing the commercial and commodified processes which contribute to the behaviours (Briggs et al., 2011a; 2011b, Chapter 8). Increasingly, I have been influenced by the work of Slavoj Žižek and the power of ideology in the context of postmodern capitalism, which, as I will show, contributes to the glittery façade to the stylistic consumption which takes place in Ibiza while, at the same time, ensures that these forms of painful excess are enjoyed *at* the expense of young working-class British tourists. Here follows a breakdown of the chapters of the book.

The structure of the book

The reader can expect to encounter various verbatim narratives and field notes from/about the Southside Crew throughout the book; this is because I use their experiences and stories to deconstruct what is taking place in the context of this cohort of Brits in Ibiza. The second chapter considers the methodology used to collect this data while the third provides a theoretical framework to my findings. The fourth chapter provides some historical and cultural context to Ibiza and how it became a global tourist destination, in particular, paying attention to the way in which the resort of San Antonio has evolved. I start the findings in Chapter 5 by considering the background of the Southside Crew and others in my sample by concentrating on what home life is like: how the 'everyday' is experienced and where and how the holiday appears on the horizon of those in my sample. Chapter 6 explores how Ibiza is constructed as a holiday destination and how the rationale to go resides in the cumulative experience gained from similar holidays around the Mediterranean as well as, more recently, a result of the rampant marketisation of the island.

Chapter 7 considers the role of identity on holiday; in particular, how British youth argue that they are 'free' within the social spaces of Ibiza by using 'anonymity' as a means for identity construction. Their *habitus* is set in further context in Chapter 8 when I offer a detailed description of how the political economy operates in the resorts, paying close attention to the role of aggressive marketing and the endorsement of excessive consumption which, in turn, results in deviant and risky activities. This provides the platform for Chapter 9 which places the deviance and risk activities of this group of working-class British youth in context through a series of case studies and field notes. Chapter 10 discusses the role of going home from the holiday and how this acts as the impetus to return to the holiday destination. The discussion and conclusion in Chapter 11 precedes the epilogue, Chapter 12, which augments the main findings from the book by presenting data from the Southside Crew in their home environment. For now, I present the research methods which prevailed in my study with this group.

2
The Flexible but Entirely Serious Methodology

> *Nathan*: Like if you don't interact with people, you don't find out stuff. Like you need to be on their level.
> *Jay*: If you had come up to us sober and asking us questions when we was minging [drunk] and stayed sober, we'd be like 'What the fuck?'
> *Nathan*: You came over out of the blue and interacted.
> *Jay*: So yeah, you come over, have a drink with us, have a laugh together. You're one of the boys now, ain't ya.

Introduction

When I approached the Southside Crew, they were drinking in a bar in San Antonio. On one hand, the fact that I approached them and frankly presented my work and its aims seemed to have meant something to them; they were obviously not intoxicated to the point that they didn't remember what my study was about, regretted consenting to participate or gave me any indication that at any point I was taking advantage of their state 'just to get data'. On the other hand, the fact that they, as well as the majority of young British tourists in San Antonio, were, to some extent, intoxicated raises a range of ethical and methodological issues about how this study was conducted. In this short chapter, I would like to reflexively engage with these issues because I see little point in denying what I did to get my data. I firstly discuss the project, its aims and briefly reflect on my sampling strategies before discussing my use of focus groups, observations, open-ended interviewing and desk-based methods. By making these discussions, I consider the reliability and validity of my data and then reflect on the limitations of my work.

The project

There were three main phases to this work.

1. Pilot focus groups and initial ethnographic fieldwork in San Antonio in July 2010
2. Continued ethnographic fieldwork in San Antonio in June 2011 and interviewing thereafter into 2012
3. Final ethnographic fieldwork in San Antonio in July 2012 and interviewing throughout 2012

The first phase of this project initially began in the summer of 2010 with some specific aims to examine the drinking attitudes among British youth abroad and explore the role of bars and clubs, and tour operators in the promotion of alcohol. However, after this fieldwork, it became clear that much of their 'drinking behaviours' needed to be considered under a wider umbrella of youth deviance and risk-taking; for along with heavy drinking came experimentation with drugs, increase in drug usage, the risk of being a victim of violence/theft, risk of rape, etc. The commercial landscape of San Antonio was also clearly playing a significant role in these practices. The following summer, my institution funded phase two, a larger study which aimed to consider these areas and a wider consideration was given to other players in the same social scenery such as casual workers (PRs, ticket sellers, bar workers, dancers), prostitutes, tour reps, club workers and the like.

On returning from Ibiza after this fieldwork excursion, there then followed a period of one-to-one, open-ended interviewing ($n = 30$) to further interrogate the emerging themes. This continued from August 2011 until June 2012, before a final fieldwork excursion made up phase three. This time, a more in-depth consideration of aspects of the local economy and commercial landscape was given. I undertook one-to-one, open-ended interviews and focus groups with hoteliers, council workers, drug experts, bar owners, police, health workers, supermarket workers, toilet cleaners, club promoters, and local residents. I also managed to broaden this sample after being contacted by local listeners of the radio station, Cadena SER, and the TV network, Televisión Ibiza Formentera, where I was interviewed. From July to August 2012, a final stage of one-to-one, open-ended interviewing ($n = 20$) took place with British tourists to validate the findings.

The sample

The book uses data from 38 focus groups with young British holidaymakers (n = 169, aged between 17 and 35); a large proportion of whom were white working class and from different areas around the UK. This work is also supported by in-depth, observational field notes (comprising over 500 pages) undertaken in San Antonio, Ibiza over the summers of 2010, 2011, and 2012. Together, this is supported by a further 50 face-to-face, open-ended interviews with British holidaymakers, drug dealers, DJs, PR workers, bar managers, and tour reps, local police, council workers, club promoters, drug traffickers and various health and ambulance workers. Informal conversations were also undertaken with these groups to corroborate emerging findings and some of the key contacts in the research also helped clarify issues of contention.

Sampling methodology and establishing rapport

For the pilot focus groups which laid the foundations for the project in phase one, young people were recruited through online social media networks on Facebook and through existing contacts. There was no real sampling agenda at this point as I was mainly concerned with establishing the research destination and some themes which would form an interview schedule. In the process of these discussions, some more precise ethical and methodological barriers became apparent. On arriving in San Antonio, the sampling methodology for focus groups was purely opportunistic and groups of young Brits were approached to participate in digitally-recorded interviews in bars, cafés, hotels, swimming pools and on beaches. The only requisite was that as many male groups were approached as female groups; that both small groups of two–three young people were considered against those travelling in larger groups (one of up to 15) were interviewed; and that younger groups in their teens were approached as well as those in their late 20s and early 30s.

I was dressed like most others in the research context to blend in with what was expected of me on holiday, wearing swimming trunks and vest most days. The holiday occasion seemed to make it quite easy to approach and be received among potential participants; most were 'up for a chat and a laugh' and considered me to 'have the best job in the world' because I was 'researching' in Ibiza. Very few groups refused to participate in the research; when they did it either was young people too hungover to even talk or was older, male groups in their late 30s upwards who, in my view, were perhaps feeling guilty about what they were doing/had done on holiday. To facilitate relations in focus groups, at times I remained neutral during drinking stories and accounts of deviance and risk-taking but, from time to time, reacted as I was surprised or shared the group's feeling about their actions.

A similar approach was also used when participants referred to violence, descriptions of sexual encounters, stereotypical views on women, and heavy drinking and drug use. This shows how I ascertained an understanding of my participants' *habitus* and how, as a result, this became important in establishing rapport. At other times, I made use of my personal life experiences in the interview/informal conversations or drew on the bank of knowledge which I was accumulating throughout the research process. A few of these people became friends via Facebook and I kept in contact with them, using their posts from time to time for my work (see Busher, 2012). Text from Facebook posts cannot be searched by the general public but to further protect the anonymity of these people, their names were changed and permission to use their posts was sought from them. Perhaps a more challenging area on the part of ethnographic researchers in contexts where intoxication is the norm is to make clear consent procedures and develop strategies to safely observe excessive drinking practices, risk-taking and drug use.

Consent procedures, confidentiality and anonymity

Quite quickly, it became apparent that it would be difficult to avoid undertaking focus groups, observations and interviews when British youth were not under the influence of some sort of substance. Indeed, in almost all the focus groups that were undertaken (during both the day and the night), young people had drunk or were drinking, and some had taken drugs. Had only non-substance using groups been approached, then the findings from the project could have been skewed. In addition, and given that the study was trying to advance current knowledge, the next step was to document what young people were doing and how they were enjoying the holiday moment and this could only be done by interviewing them in these specific situations of consumption. Of course, some note the problems of this approach in that participants may be overly compliant in agreeing to be interviewed (Measham et al., 2001) but no pressure was made on the people in this book to contribute their experiences and opinions; if anything, the contrary as most groups were enthusiastic about the project. Every group that was approached for recorded interviews was aware of my research intentions, saw examples of the questions which were to be asked and thereafter, gave consent to participate.

To avoid formalising the research process and facilitate rapport, the aims of the project and consent procedures were presented verbally on the digital recorder and consent was given this way (see Briggs, 2010; 2012; Treadwell et al., 2012). Secondly, in the original proposal, attached to the consent form was the term 'binge drinking' which, after conducting some of the fieldwork in phase one, seemed to generate some animosity among a few young people because it was almost assuming that was what they were doing. Although this may have been the case according to established definitions (Kuntsche et al., 2004; DHS, 2008), the term is also loaded with negative stereotyping largely thanks to the antisocial connotations the government have produced around the people who allegedly do it. Thereafter, 'binge drinking' was substituted with 'heavy drinking'. Lastly every group was given the opportunity to listen to their interviews and were given out contact details should they want to request information on what they said. All data remained confidential and was not disclosed to a third party without the prior agreement of the participants themselves and this was assured through a signed consent form. Furthermore, all participants, as well as particular local bars and clubs, were also anonymised.

To drink or not to drink: That is the question ... so what is the answer?

Undertaking ethnographic research often means accommodating the participants under study, or even in some cases, radically changing the research direction to adapt to particular practices and/or the social context of their behaviours to better the research endeavour. To advance current knowledge in this particular subject area – that is to go beyond surveys and understand the subjectivities attached to the deviant and risky behaviours of British tourists on holiday as well as the commercial pressures which influence them – I drank alcohol and went to certain places with the research cohort (bars, nightclubs, brothels, strip clubs). Of course there is no correct way to research these issues and my approach has benefits as it does drawbacks. For example, while on one hand, by drinking with my cohort, I walk into an ethical and methodological no-man's land, however, on the other, I ask myself how much can people remember they drank or how many drugs they took when they complete a survey after a week of non-stop intoxication in Ibiza?

In my mind, there therefore seems to be no perfect means of data collection without its downfalls.

Initially, it was stated in the ethics proposal that drinking alcohol would take place in order to some extent facilitate relationships – otherwise I would have looked out of place; especially in a social arena where the general expectation is to drink or get drunk. However, in Ibiza I realised that when with groups it was beneficial to 'keep up' to some degree with their drinking pace, since they were engaged in group drinking and I had immersed myself in their space. As the quote indicates at the start of the chapter, drinking alcohol with people like the Southside Crew helped me build relations with them. Had I not, then I feel it would have been detrimental to the quality of the data I retained. In this respect, I am not alone in this approach (Measham et al., 2001; Andrews, 2005; Blackman, 2007). I didn't coerce my participants into drinking more than they should and only 'went along' with what was expected of me under the specific group dynamics and social moment.

Therefore the challenge I faced was how to manage relations and make decisions with drunken participants in such intoxicated contexts. In the situation of an ongoing carnival of drinking, which does not tolerate hierarchical difference between social groups, staying outside the common practice of drinking means either to be excluded from the celebrating group or put at risk, if not destroy, the event itself. A non-drinker, a person who is not drinking or who refuses to do certain stupid things when the others are 'up for it', can be downgraded in his or her status and even excluded from the group and this rule applies to ethnographic researchers as well. By drinking with participants, I stood a greater chance of documenting practices and attitudes closer to those held by members of a group and this had its benefits.

I acknowledge that I may have collected data in which participants were drinking and/or were reflecting on drunken events (Griffin et al., 2009) and their accounts may not have been entirely accurate; however, this is the nature of the social occasion and this is 'what happens' for British tourists on holiday (Briggs et al., 2011a; 2011b). I therefore see it as my job as a critical realist – someone acknowledging his/her presence and interactions and how they contribute to the co-production of social meanings – to unpack these experiences and events as much as possible. To do so, it was clear that I had to become part of the interactions (Coffey, 1999). As Patton (1990: 474) notes, ethnographic researchers 'should strive neither to overestimate nor to underestimate their effect [on the research study] but to take seriously their responsibility to describe and study what those effects are'.

Focus groups

Because many of these tourists travelled with friends (and friends of friends), I felt it was best to undertake focus groups rather than select people individually and coax them away from their group for one-to-one interviews. In this respect, focus groups also enabled me to capture the 'group moment' and the experience of deviance and risk in the holiday context. The focus groups were open-ended which enabled groups to determine how people talked about or perceived various aspects of their substance use and risk behaviours (Carlson et al., 1994; Griffin et al., 2009). In some of the focus groups in phase one, interviewing groups of young people in pairs was useful (Raby, 2010), and one researcher took the lead with questioning while the other remained in the background, offering strategic direction at intermittent periods. This deterred us from dominating group dynamics and served to empower participants. Undertaking focus groups meant it was difficult to disaggregate individual feelings on particular issues; especially when the group were mostly used to talking about substances and sex and there was a certain element of bravado in the group context (more male than female). Moreover, not every member of the group contributed to the same degree in conversations.

When this strategy was managed well and the groups were talking about themes familiar to them individually as well as collectively (Warr, 2005), then it was possible to gather what may be considered to be sensitive information. In focus groups, as well as during observation periods, it was important to utilise episodes of acting-out, or presenting a particular image in the presence of others. Indeed, Hyde et al. (2005) show this can be highly revealing when attempting to understand the normative rules embedded in the culture from which participants are drawn. This meant passing non-judgemental comments, and, at times, laughing along with or appearing impressed with raw accounts of sex and substance use.

Observations

I conducted observations in bars, clubs, beaches, and general touristic areas in San Antonio and to a lesser extent Platja d'en Bossa. Objectively – that is without interfering with what was taking place – I witnessed illicit behaviour such as drug use, drug dealing, fighting and violence, self-injury (and consequences of it) and general deviance and risk-taking behaviours. However, when drinking with the Southside

Crew and other British tourists in bars up and down the San Antonio Bay and on the West End, I was perhaps more 'participant' than 'observer'. But because I was interested in understanding the subjective intentions this group of tourists ascribe to the holiday occasion against the macro, socio-structural forces of commodification, there could have been no other way. As Meethan (2001: 162) notes 'If we are to account for localised change, then rigorous micro-ethnographic techniques will need to be employed as much as macro analyses of the global context' and this is precisely what I have attempted. It was this 'inside experience' which allowed me to document aspects of the holiday and its hype (Chapter 5), constructions of the holiday (Chapter 6), image and identity (Chapter 7) against exposure to the commercial and commodified pressures of the San Antonio landscape (Chapters 4 and 8). Taken together, this gave a detailed and nuanced picture of how moments of deviance and risk-taking evolved in the social context (Chapter 9) but also why so many British youth consider it so important to come back to the same place to do the same thing, year after year (Chapters 10 and 12).

A delicate combination of overt and covert roles was used because I was not able to disclose my research intentions to all (Briggs, 2012) and this was particularly the case when undertaking informal conversations. This type of data collection was reserved for PR workers who often dealt drugs, strippers and/or lapdancers and with other very intoxicated British youth (who were too drunk/high to consent to recorded interviews). Making observations was not always easy, especially in places like the San Antonio West End, which is a very concentrated space made up of bars, clubs, fast food outlets and strip clubs (see Figure 2.1). During the day, it is reasonably tranquil but at night it becomes immensely chaotic when numerous single-sex groups of mainly British tourists collide between numerous PRs, prostitutes, lapdancers and strippers. In this narrow, concentrated space designed for consumption and excess (Miles, 2010), the 'data-gathering' moments are never-ending and it becomes difficult to work out how and what to document. Nevertheless, I just did my best to record what I could using my BlackBerry phone.

Unsurprisingly, some of the fieldwork days were extremely intensive; lasting up to 18 hours a day and it was paramount that field notes were written up as quickly as possible. The field notes resulting from these observations and informal conversations were converted to detailed descriptions of participants and their activities in an effort to provide a *'written representation of a culture'* (Van Maanen, 1988: 1) or 'lived experience' (Geertz, 1973), although, at times, they were autoethnographic

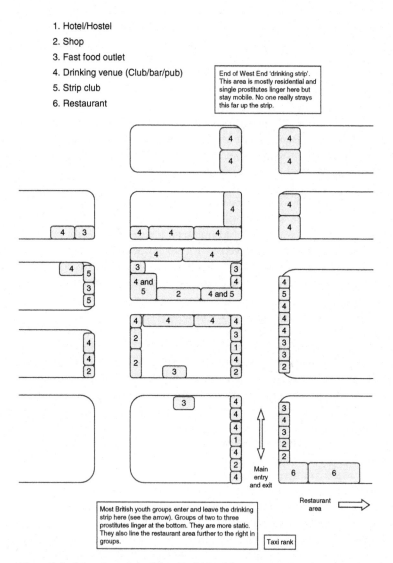

Figure 2.1 Mapping of the West End 'drinking strip' in San Antonio, Ibiza
Note: (1) Hotel/hostel, (2) shop, (3) fast food outlet, (4) drinking venue (club/bar/pub), (5) strip club and (6) restaurant.

in nature (Ellis and Bochner, 2000), providing a cultural connection between myself and others in the same space.[1]

Open-ended interviewing

Themes from emerging findings which were generated from phases one and two of the field trips in Ibiza provided the topics for open-ended interviews ($n = 50$) with various players of the social scenery. These included 'experienced Ibiza goers', individual members of British holiday groups, drug dealers, DJs, PR workers, bar managers, and tour reps, local police, council workers, club promoters, drug traffickers and various health and ambulance workers.[2] Some of these people were interviewed in the UK, in Ibiza and on the telephone where no meeting could take place. This also enhanced the reliability and validity of the data as well as augmented existing knowledge and helped situate atypical cases (Briggs, 2012).

Data analysis

Once transcribed, interview and observation data were categorised thematically from phases one and two, with the key areas of investigation providing the overall framework for coding (Ritchie and Spencer, 2004). Analyses were inductive, which meant that themes emerged from the data rather than being hypothesised. These initial analyses laid the foundation for making valid (or otherwise) the findings from the study when commencing phase three of the study which predominantly relied on open-ended interviewing.

Limitations of my work

There are a number of limitations to the work I present here. Firstly, I have not sought to establish long-term or shifting patterns of substance use, deviance and risk-taking among specific groups of British youth. This, to some extent, has already been done (Bellis et al., 2000; 2003; Hughes et al., 2004; 2009). I am more concerned with how and why moments of deviance and risk-taking arise and under what conditions do they flourish. Secondly, I acknowledge that my epistemology lies principally with the *habitus* of this group of working-class British tourists in only one resort location on one holiday island. However, I believe that much of what I have found resounds with what takes place in other resorts around the Mediterranean among British tourists (O'Reilly,

2000; Andrews, 2005) and other groups from other countries (Tutenges, 2010; Redmon, 2003), there are specific social conditions and structures which encompass Ibiza – for example, drug use is more prevalent, and Ibiza harbours the Superclubs and very elite-like nightlife venues – which are not found to the same degree elsewhere.

In this work, I have not given full consideration of gender relations (see for example Skeggs, 2004) in the holiday context, largely because this has been done with British tourists (Andrews et al., 2007) but also because I was more focused on how and why deviant and risk moments arise on holiday. There is, however, some acknowledgement of the gendered holiday space, especially the objectification of women as commodities (Chapter 8). Relatedly, in my discussions with young men in this study, there was the danger that they might jovially upgrade their accounts of drinking, drug-taking, sexual encounters and risk-taking to receive specific social kudos from the group (Chapter 6). This was one reason why I sought to follow groups such as the Southside Crew on their nights out and throughout the holiday to see if what they were saying related to what they did/how they felt in practice. Where possible, I tried to unpack this bravado and locate, as closely as possible, the source of 'what happened'. This was often a painful and slow process, especially in focus groups where a continual deconstruction of events often prompted some of my interviewees to get impatient.

Conclusion

To undertake this study, I had to redefine some ethical and methodological issues concerning the explicit behaviours under observation as well as its social context. I did these things because I was interested in advancing the field and taking forward an enquiry past the established realms of epidemiology into a new subjective territory which was more attuned to contextualising micro socio-interactional dynamics against socio-structural forces. In the next chapter, I situate the findings of my study in a theoretical framework.

3
The Theoretical Framework for the Study

Introduction

The primary function of this chapter is to set in context the main empirical findings for the book by utilising a theoretical framework. The disciplines from which those theories derive are Sociology, Criminology, Tourism Studies, Leisure Studies, and Economics, and take predominantly a critical standpoint on the recent changes which have shaped the world. I feel increasingly that we cannot ignore some of these disciplines in our analysis of the postmodern world and should not concentrate their use so independently in our work. If you recall from Chapter 1, I was keen to avoid already-established pathologising exercises on people like the Southside Crew so the theories I have chosen should give you the basis to understand the findings on a number of levels: historical, structural, social, cultural, spatial and subjective. The key questions I will come to answer in the pages which follow but here I am interested in setting context to some of my questions: Why do this particular social group have these attitudes to their holiday? What puts these people in these situations in this particular time? Why might they engage in these behaviours? The theories I discuss here are the principle ones which provide this basis, although throughout the book, I will also reference other theories where necessary. I begin by briefly charting the main structural, social and cultural changes which have affected the world over the past 30 years before discussing in detail how this has shaped both the collective and subjective experiences of the people you read about in my book.

The postmodern world

From the mid 1970s onwards, the world advanced into new and uncharted territory: into one of uncertainty, risk and precariousness.

It was from this period that an increasing global, neoliberal agenda – based in principle on free-market economics and individualism – led to the dismantling of the post-war golden age of high employment, welfare, stable family structures, and consensual norms and attitudes, leaving in its place, market and economic instability, 'risk', potential ecological catastrophe, high unemployment, and increasing social inequality sharpened by welfare reduction and a criminalisation of the 'underclass' (Lea, 2002; Hayward, 2004). These are neoliberal economics which are concerned with deregulation and removing perceived restraints which impede profit, yet it has with it brought new uncertainties and inequities. This move towards a 'risk society' (Giddens, 1991; Beck, 1992) is important because it has resulted from the downsizing of traditional industries, the dominance of global markets and rising unemployment and the growth of low-paid work (Crawshaw, 2011).

In this increasingly globalised world, where populations are displaced for various political, social and economic reasons, competitiveness ensues and there is a constant battle among corporations to draw profit regardless of social consequences; which means farming out production to countries where a blind eye can be turned to human rights. The rest of us? Well we consume their misery. The transformation is thus one from production to consumption. In this process, argue Ferrell et al. (2008: 14), this happens:

> Late capitalism markets lifestyles, employing an advertising machinery that sells need, affect and affiliation as much as the material products themselves. It runs on service economies, economies that package privilege and manufacture experiences of imagined indulgence. Even the material fodder for all this – the cheap appliances and seasonal fashions – emerges from a global gulag of factories kept behind ideologies of free trade and economic opportunity.

In this world people are left to make 'free will' choices which should enable them to be 'good citizens', to make the right decisions, find the right jobs, partner and make a successful life which is grounded in all its symbolisms: good job, house, car, nice clothes – these are the end goals which are culturally assimilated for us to achieve: they are consumer items. The main challenge to the realisation of these material dreams is that the closer to the rainbow people think they get, the further they have to travel to find its source. Really it is an endless journey: one which seems to stretch further as they move through the life project. So as the horizon moves, so too does the rainbow because people end up always wanting the 'next best thing' which is marketed to them. Indeed, in the climate of increased consumption, swathes of new industries have expanded 'choice' available to us while, at the same time, opened up a world of leisure and consumption to fill the individual's sense of *being* generated by this structural detachment.

Postmodern identities

From this period, state and community retraction coupled with market permeation have created a separation from established and traditional social values grounded in stable employment, family and community. It is this decoupling which produces a subjective existential impasse for many people today: a crisis of who they are, a crisis *being*, known in academic terms as 'ontological insecurity' (Giddens, 1991; Beck, 1992). Without the past social determinants to direct humans into what they should *be* and what is expected of them, in the neoliberal era people are instead left to their own devices; they are responsible for themselves. Very often, it is this individualism which promotes a self-love which sidelines collectivity without thought of the other: a sort of *'me against the world'* divisiveness prevails or, if we were to extend a familiar Apple marketing strategy, 'Ipod, Ipad, Iworld, I'mtheonlyonewhofuckingmatters'. It is this individualised nature of being which has almost narcissistic implications for attitudes to the body and image; where health, fitness and 'youthfulness' and beauty take a particular moral form (Bunton et al., 2011). And this is precisely it: people's values these days tend less to reside in the conceptions of work, family and community but in elastic, individualised identities about the self and the body which are constructed and moulded by consumer lifestyles, the online world and the media.

However, at the same time, people must navigate this uncertain territory and, in the process, make something of themselves. After all, the superstructure which is neoliberal capitalism endorses such a life project – one of 'you can be who you want to be' if only you can realise the means to do it. If people fail, the pressure may haunt them and this lack does not easily disappear; it is for life (Bauman, 2007). Related to this as a consequence of this disembeddedness is reflexivity: the capacity to reflect on individual decisions/actions and the potential suffering/guilt attached to them. We are therefore talking about a process of 'reflexive modernisation' which prompts the construction of the self by reflexive means (Beck and Beck-Gernsheim, 2002). Much like Prometheus's liver which replenished each day only for an eagle to gauge it out, the *'what ifs'* and *'should haves'* of life live on in the everyday psyche of the individual: they may temporarily evaporate but they soon return. In this process, personal narratives are reflected upon for their value and subjectively superimposed against the hegemonic cultural framework for validation which we have determined thus far is participation in consumer lifestyles. One simplistic example of this may be a young man reflecting on his inability to get a job, and in doing so, considering himself somewhat inferior to his friends because they have work, and as a consequence, can participate in the dominant social order by buying a car, having nice clothes, going on holiday to a nice destination, etc. Of course there are other elements to consider such as class and place but my point here is that individual shame is often self-constructed as a consequence of failing to make the grade in social terms.

In a postmodern context, because identity is malleable, it is not retrieved or gained. People convince themselves through mantras like 'I came here/did/have this and have, in the process, found myself' that they have somehow managed to assemble who they are, thereby understanding their inner core. But these days identity is pliable to social situation, context and space, which combined with its reflexive power, increasingly results in 'reinventions'. The self never arrives at its destination because it never *becomes*. The result is that people often feel incomplete and inadequate. The commercial world has capitalised on these personal insecurities evident in the abundance of industries around beauty, leisure, and tourism which promote consumption and the 'next best look', the 'latest edition' or 'best place to holiday'. Thus the main practice of *being* these days for people in my sample – to retain ontological security, feel accomplished and included – is through participation in these industries which collectively represent capitalist consumer culture.

The decline of work-based identities, and the growth of consumer culture and leisure pursuits

Once upon a time, in the industrial periods, many of the people in my sample would be bound to quite stereotypical class-based conceptions of work and family life. However, as I noted earlier in this chapter, times have changed and the concept of permanent jobs and secure family life have liquidised and been replaced by a world of temporary contracts, part-time work and unemployment. This is particularly the case for working-class populations in the UK. Perhaps as a reflection of rampant individualism championed by the Thatcher and Reagan years, today's generation, while resenting the insecurity of the world, want to 'be their own bosses' (Standing, 2011). Yet these preferences are generally unavailable to people in my cohort because they often enter jobs with little employability, receive low wages and fewer benefits. Entering at the bottom of the ladder in something temporary or part time limits upward social mobility because permanent job opportunities are disappearing. In addition, the work is generally neither productive or enjoyable; it is 'careerless' (Standing, 2011). As a consequence, many people in this bracket lack work-based identities and this generates actions and attitudes towards 'opportunism' and 'seizing the moment' because there is no certainty of the future work or career to guide their commitment. Even those in work find themselves 'spent' and find little other motivation than to engage in passive forms of play; the means by which they enjoy themselves and seek activities to do so is reduced (ibid., 2011). Hardly surprising perhaps then that there is quest to transgress when even institutionalised situations like the holiday easily trump those of workday existences (Presdee, 2000; O'Neill and Seal, 2012).

It is into this void of work-based identity (Rojek, 2005) that steps the role of 'play', leisure and consumer culture as a means of self-realisation (Schor, 1992; Smith, 2012). Traditional 'leisure careers', which would normally run parallel alongside stabilised labour forms among the working class now look increasingly redundant and instead virtual lives (Standing, 2011), shopping and participation in the night-time economy (NTE) (Hall and Winlow, 2005a) and in some cases, criminality and drug use fill the time and are the primary means for status generation (MacDonald and Shildrick, 2007; Hallsworth, 2005). Surplus money is therefore normally spent on home improvement, luxury commodities and exotic holidays thereby acting as some proof of the addiction we have to consumption (Schor and Holt, 2000).

Moreover the 'subjective' experience of the individual is increasingly artificially constructed through 'group approval' (Standing, 2011) which is perpetuated by elements of cultural competition – people feeling they need the new trainers, the next designer items, and new car to succeed in the battle for social recognition (Hall et al., 2008), and to attain positions of social distinction (Bourdieu, 1984). Such an emphasis on status-placing activity as a means of identity construction is not a new concept. Over 100 years ago, Veblen (1994) posited that it was the wealthy *leisure class* who obtained this status through their access to economic capital and displayed the capacity to spend it willy-nilly. Veblen called this *conspicuous consumption* or the extravagant display of wealth through gratuitous forms of expenditure, which in turn, exhibited status superiority and it was this display which often acted as the guide for those in the lower class brackets as a form of status emulation (Corrigan, 2010). Veblen argued that participation in these social circles meant displaying cultural criteria and codes of behaviour; however, importantly, he recognised that members of this class perhaps didn't have the 'freedom' they thought they had because they were bound to this form of coded leisure and if they strayed their cultural standing would be at risk.

These days, however, this cultural framework is far more advanced. Products and commodities are branded; they are themed accordingly with symbols to increase consumer desire for them. So the growth of consumption as a mode of claiming status (Veblen, 1994) has 'placed the mass of people in a position of having more choices than ever over their objects and ways of doing things and spending time' (Corrigan, 2010: 84). Moreover, representation of leisure and consumption come through new cultural mediums such as television, film, advertising, music, and this, to some extent, organises our ideals of gender, legitimacy, social value and responsibility (see Hayward, 2004). Spaces, like for example shopping malls, airports and holiday resorts are made commercially defined so they can operate as containers for human interaction and consumption (Miles, 2010). Even people are branded and themed! For example, David Beckham and Madonna function as role models and identity objects that shape our leisure behaviour and consumer practices; they both have branded commodities attached to their name as well as other commercial ventures. Leisure now is then a site for self-identity, lifestyle and stylish consumption (Featherstone, 1990; Corrigan, 2010). It is a means of 'casual leisure', seized opportunistically and motivated by immediate gratification through activities such as drinking, smoking, and shopping (Rojek, 2000). As Hayward (2004: 85) argues, it is the:

> Unique feature of an unmediated consumer culture is the extent to which it propagates [within] individuals the constant demand for more – more products, more excitement, more stimulation, more experiences.

This has had substantial implications for the working classes who once had secure jobs and increasingly have to survive on the periphery in this structural vortex of temporal and 'careerless work'. Successful navigation of the precarious labour market can reap benefits in that people can move from job to job relatively easily regardless of the previously established barriers of patriarchy and family. There is on the horizon for these people the green meadows of consumer culture in the pasture foreground. They are marketed lifestyles beyond their means and are in a perpetual struggle to keep up with the new trends of the elite (Veblen, 1994), who, in turn, distance themselves from the mass availability of products through high culture, the cultivation of taste, art and aesthetic styles. They seek to distinguish themselves from this naked materialism. However, for those who cannot realise these dreams, a heavier deprivation potentially arises: one lack attached to the failure of market participation – 'looking good'. Now such a life outlook may have its social credence in that people admire an image: someone who can project wealth and success and is a clear participator in the expectations of materialistic being. There is perhaps no better place to see these processes at play than in the NTE, in the clubs, bars and pubs in town centres across the UK, where club cultures, new attitudes to intoxication, deviance and risk-taking have also evolved over the last 30 years.

Clubbing, the culture of intoxication, and deviance and risk-taking

It was claimed that a rise in disposable leisure time gave rise to 'club culture' in the 1980s which dissolved structural divisions such as class, race and gender as the dance floor crowd became collectively immersed (Redhead, 1993). Rave and dance music was said to represent the demise of the subcultural oppositional youth styles of the 1960s and 1970s. As Redhead (1990: 2) put it, the subcultural struggle was redundant because young people and the rave experience was 'more resonant of shopping and consumption rather than resistance and deviance'. And although there have been recent efforts to suggest that clubs represent spaces for universal expression of 'togetherness', fun, relaxation and pleasure and contexts of temporal association of club crowds or neo-tribe (Maffesoli, 1996; Malbon, 1999), we have seen an increasing fragmentation of culture which renders the concept of subculture problematic. As Chaney (2004: 47) notes 'the once accepted distinction between "sub" and "dominant" culture can no longer be said to hold true in a world where the so-called dominant culture has fragmented into a plurality of lifestyle sensibilities and preferences'. Instead terms like 'lifestyle' over 'subculture' have been favoured because of its focus on consumer creativity; that is how young people construct and reconstruct their image and identity through ongoing shifts in musical or/and stylistic taste (Miles, 2000).

Indeed, the last 20 years have seen a rampant increase in the popularity of clubbing, raves and general participation in the NTE in town centres and cities across the UK. Partly buoyed by the disintegration of the town centre, new marketing and advertising strategies have reinvented this space around NTE consumption (Hobbs et al., 2003; Hollands, 2002). Traditional pubs have been replaced by swanky bar/bistros and 'happy hours' have changed the nature of alcohol consumption for many in my sample. As Hayward and Hobbs (2007: 443) note the 'liminalisation of space within the NTE should not be understood as "spontaneous manifestations of the carnivalesque", primarily because of the extent of the influence of big business in structuring the desires and attitudes of young drinkers as well as the environment in which they are realised'. It is in this context that weekends have become a popular bracket whereby these forms of consumption take place while, at the same time, acting as a temporal departure from routine and repression. They are, as Blackshaw (2003) argues, the occasion for ritualised types of leisure practice that celebrate excess, physicality and the suspension of responsibility which is why leisure ritually licences forms of transgression such as drunkenness, taking drugs and violence. And while this form of intoxication is nothing new (Maffesoli, 1996; Bakhtin, 1984), in this book, I am interested in the way in which these capital forces take the lead and manage these rituals.

For working-class groups, it is this occasion (the weekend) and these spaces (the town centre NTE) which reaffirm a sense of belonging, reflect a sense of collective identity as well as gender relations; such as guys staring at girls and 'pulling' them as a means of reaffirming working-class masculine youth group solidarity (see Winlow, 2001; Rojek, 2005). Therefore these spaces offer the means for performatively sustaining masculine identity. However, the struggle for an identity and lifestyle in these arenas, which is measured by social distinction (Bourdieu, 1984) in a society of increasing class divisions, instead reflects a sense of powerlessness attached to cultural and economic exclusion coupled with a depoliticisation of class consciousness – of who this group *are* collectively as well as individually; and this is reflected in what they do on the weekends in the UK:

> Inequalities in access to economic resources and differences in prestige directly influence leisure forms and practice. Thus, masculine leisure repertoires in working-class youth culture of excessive drinking, having a laugh, occasional outbursts of violence and generally denigrating the system is an on location response to formidable cultural

and economic inequalities of class. Working-class men and women are positioned to follow these scripts of leisure behaviour that derive from their relative powerlessness. Although they involve rich forms of creativity and resistance, and constitute important parts of social solidarity, essentially they are fatalistic responses to their class isolation from decisive control and influence over property and knowledge. (Rojek, 2005: 149)

Indeed, part of this cultural inventory is the growth in cultural acceptance of intoxication and recognition that young people like to get drunk and high; it is now a context of normalisation of drugs (Parker et al., 2002) and/or 'determined drunkenness', of complex drug and alcohol repertoires (Hadfield and Measham, 2009) which forms part of a broader acceptance of the cultural context of risk-taking (Measham, 2006; Wilson, 2006; Measham, 2008) reinforced by popular culture in TV dramas, films, and public disclosures of celebrity lifestyles (Blackman, 2011). In this way, risk linked to intoxication has become a normal and celebrated feature of mainstream contemporary leisure. Or as (Ferrell et al., 2008: 72) say:

> In this context, transgression can be seen as a breaking of restraints, an illicit realisation of immediacy, a reassertion of ontology and identity ... where identity and emotion are woven into the experience of rule breaking.

In this world, the immediate is given preference and hedonism is a means of self-actualisation (Hayward, 2004) – of becoming and *being*. They are carnivalesque moments where life and rules are turned upside down as 'time out of time' is claimed (Bakhtin, 1984; Redmon, 2003) as well as the emotional release and pleasure from rule breaking (Katz, 1988; Presdee, 2000). Bakhtin (1984: 122–3) notes that the most significant elements of the carnival (low behaviour or transgression) are the dissolution of the everyday *'hierarchical structure'* of society in favour of 'free and familiar contact among people' which means different people interact in different ways than they would under the normal conditions of everyday life (high behaviour or conformity). Featherstone (1991: 22) describes the carnival as:

> The popular tradition of carnivals, fairs and festivals provided symbolic inversions and transgressions of the official 'civilized' culture and favoured excitement, uncontrolled emotions and the direct and

vulgar grotesque bodily pleasures of fattening food, intoxicating drink and sexual promiscuity.

It can signify an attempt at 'freedom', a moment of self-realisation which can offer social stability, and credibility because the behaviours are made with the potential for future storytelling (Briggs, 2012; Griffin et al., 2009); something which declares some proof that their excesses symbolise to others they are not 'boring people' (Tutenges, 2010). Some say that transgression and testing the boundaries are a natural part of the life course (Maffesoli, 1996; Tutenges, 2010), but these days such is this commitment to 'enjoy' that very often engaging in play often breaks the boundaries of pleasure into realms of pain or *jouissance* (Lacan, 2006); that is, it becomes more 'normal' – regardless of how painful it is – to seize these moments of pleasure, risk-taking excess and transgression than not to (see Hall et al., 2008). This, in particular, has specific relevance to what I have to say about attitudes to deviance and risk-taking on holiday.

Ideology

Why is it therefore that people would put themselves through these things to retain some sort of ontological stability? What is driving this, what appears to be from the outside, inherent determination to almost self-destruct at weekends and on holiday? Well, because the kind of lifestyles, values and attitudes I am describing among the working classes are situated under structures of economic, cultural and political inequality, it becomes crucial that ideology masks how power works to advance established political and economic interests (Rojek, 2005). In the current context of the global social order – which is consumer capitalism – this is to persuade people to participate in the working world so they can consume, spend, and as a consequence, retrieve a sense of 'happiness' and fulfilment from that. And by doing so the very same global machinery is sustained. In fact, Braham and Wagg (2011: 2) put it perfectly when they say that:

> The ideological tasks for modernity are both mediation and diversion: to encourage workers to acquiesce in capitalistic work discipline and exploitation, and to convince them that their future happiness would best be secured as willing consumers rather than as dissatisfied socialists and radical class warriors.

This is why these days, as Ritzer (2000) points out convincingly in his McDonaldisation thesis, everyday life is branded and increasingly multinational companies have influence over the contexts in which leisure and lifestyle choices are made. In this way, it is the prominence of the corporation and the State which formulate commercial and bureaucratic controls that produce mass conformity. Putting it more bluntly, the choices people make are less 'free' than they would like to think because they are made within a structuro-culturo framework of consumer culture:

> The development of consumer culture takes capitalism into a new phase characterised by a more sophisticated form of domination over the masses. Under it, control is accomplished through 'free choices' made by consumers in leisure and consumption activity. These 'choices' are directed by the culture industry to achieve conformity, docility and the reproduction of capitalist hegemony. (Rojek, 2005: 72)

And it is because ideology assumes this power that much of what the people in this book say, feel or think is often governed by this framework – one which wants to feel 'free', while at the same time, remain powerless. As Žižek (2011: 359) says:

> The lesson to be learned is that freedom of choice operates only when a complex network of legal, educational, ethical, economic and other conditions form an invisible thick background to the exercise of our freedom.

Or, as I wrote in a poem, what I am saying is that:

> All this anxiety,
> Is about too much variety,
> Because a marketed choice,
> Makes passive our voice;
> Makes us docile, makes us numb.
> Because what is available,
> Is to what we succumb.
> It is what we *become*.
> Thus a commercial demon,
> Robs us of freedom,
> Leaving unhealthy our sobriety,
> Among manmade temptations,
> In a consumer society.

In the context of excessive consumption, it may be that people consider that engaging in hedonism is what makes life worth living but that they are instead undertaken as part of the 'artificial colouring service to emphasise the greyness of social reality' (Žižek, 2011: 9) – to construct something real from the unreal and false experience of consumer capitalism (Žižek, 2008). In fact, some of these issues are somewhat recognised by the Southside Crew. In this jumbled exchange while drinking in a pub one night on Southside, the following issues arise: some obvious discomfort about the incessant pressures consumption promotes as a form of *being*, feeling 'trapped' as well as angry towards a socio-political system which they seem to see as restrictive and unfair:

Nathan: We were not meant to be domesticated animals.
Jay: Everyone wants to be the best.
Nathan: That's what I'm saying, mate. We are not meant to be domesticated animals, we are meant to be free.
Jay: Everyone these days wants everything, fucking Ferrari, fucking top boy. This is the way we are brought up.
Nathan: It's to do with politics, mate. We were not supposed to be designed to live like this. We have no freedom; we are in prison, I can see the bars. We are fucking slaves, mate. All politics [starts to get angry]. In Africa there is genocide going on yet they [the British army] are fucking going to Iraq, fucking Afghanistan, fucking oil.
Jay: We are wasting time in Iraq. Billions of pounds. Same shit will happen, we get the oil and then some bullshit dictator will take over and then we have to step in again.
Nathan: It's the big fucking oil companies telling them [politicians] to go and get it. We are the evil empire. We are fucking evil.
Jay: This country is fucking shit, mate.
Nathan: We fucking murdered, raped, did shit to all sorts of cultures. This country is fucked.
Jay: The government is one big gangster; they take all the money they want to take. All they want to do is make money.
Nathan: That's when, when the riots [across England in 2011] was on I was fucking lovin' it. I wanted to see them go down [the government]. They hide behind the law, no one sees the truth.
Dan: But you do know what is going on. Listen to you. You know it.
Jay: But there is nothing you can do about it ... the law is fucked up.

The helplessness of the situation captured perfectly in Jay's last words: there is nothing people can do as individuals apart from live by the hegemonic cultural ideology which fronts this unequal and restrictive system.

Conclusion

The chapter has aimed to set the theoretical context for my book. I have chosen these perspectives because people in my sample are predominantly working class, white and are familiar with drinking at weekends, taking drugs, and say they have generally 'boring' jobs and/ or are studying, or don't see much prospect. What I am trying to say is that few see their work life as important or valuable and instead express a commitment to leisure pursuits and 'the weekend' and 'time out' for themselves such as the holiday (Chapter 5). They are also used to engaging in forms of deviance and risk as a means of identity construction as some form of 'real' experience from the unreal materialism which ideologically governs their ontology. In the context of this book, I am therefore interested in how such a social system has weaved its way into the subjective individual politics of this group of working-class British people and is strategically yet subliminally guiding them in their life projects through the offer of endless commercial and commodified nothings in situations and places designed to leave some impressionable proof on their psyche that they have led 'fulfilling lives'. And one epitome of this process is evident in what is currently taking place in San Antonio, Ibiza, and the next chapter considers how this location has evolved to facilitate these features of postmodern life to people like the Southside Crew.

4
Ibiza: The Research Context

Introduction

For centuries, Ibiza was a trading outpost and people survived in poverty and isolation. Now it stands as one of Europe's highest rankings for per capita income for economic growth and tourism. It's clear that since the 1960s, tourism in Ibiza has transformed the lifestyle, living standards and opportunities for many of the island's residents as well as provided it with an economy. Commercial enterprises, global chains, Superclub shops and branding and the general commodification of anything remotely touristic is commonplace across the island. While tourism now makes up approximately 80 per cent of the GDP and 4.5 million people visit annually, as we have seen, with it come significant problems of crime, drugs and disorder, which is a major burden to the island's infrastructure. This reputation has, and in the words of the hoteliers and local business owners, impacted on the reputation of San Antonio and, to some extent, Ibiza. In this short chapter, I would like to introduce the reader to San Antonio and chart the main changes to the resort through the testimonies of those who have witnessed first-hand its commercial evolution.

Destination San Antonio

From Ibiza airport, San Antonio is about 20km and there are two routes: one which takes you through the heart of the island past the Superclubs, Amnesia and Privilege (main route); and the other, a slightly longer route which passes through some quiet Spanish villages. At any time of day or night, the main route through the island sees numerous taxis passing through at great speeds, shipping tourists to and from the airport and/or the Superclubs. Indeed, in the evening, driving over the hill at Sant Rafael next to Amnesia, a distinct glow appears on the horizon; this is the nightlife hub of San Antonio. And as one approaches, it is like driving into an adult adventure playground.

The main road into the resort passes numerous hotels, which, as you draw closer to the seaside and bars and clubs, seem to get bigger: the main one being Es Pla, a notorious Club 18-30 package tourist resort for British tourists. By the time you reach the roundabout (*'el huevo'* or 'the egg' as the Spanish call it) which divides the route to the left along the bay or right into the town, the party is well under way for most people. To the left is San Antonio Bay, which starts off as a sort of dirty, sandy beach on the right with the main clubs Es Paradis and Eden opposite, a fairground and various bars which, as the bay disappears, become more sparse on the left. The bars continue but generally play more relaxed music and are often a stop-off for groups of British tourists heading into San Antonio for a night out on the West End. The cheaper hotels, which are more popular with British families and elderly tourists, tend to be scattered up this coastline around the Bay and around other desolate and derelict areas where half-built hotels await completion.

To the right of el huevo are various bars and restaurants which sell popular beers, English breakfasts and numerous tourist shops which sell items such as topless postcards, tabloid newspapers and sun cream as well as other beach accessories. Walking up here, you encounter the marina on the left where depart various family boat trips as well as 'booze cruises' and to the right, between the fountains and shaded sitting areas of the town's *plaza* (main square), are numerous bars and restaurants. The nearer the West End, the greater the concentration of the bars and restaurants, and by the time you arrive at the bottom of the 'drinking strip', there is nothing other than bars, clubs, strip clubs, and take-away outlets. This strip is composed of a narrow road with several accompanying roads with similar modes of consumption to the side. Around this area are cheap hotels, hostels, pensions, as well as cafés and small local supermarkets. At night, the main West End area swells with predominantly British tourists. My field notes here describe the experience walking up and down the main strip at the height of the mayhem:

> It is 2 a.m. when we leave the hostel. As we walk to the strip, we pass numerous drunken Brits; one of which sits retching over the kerb with his mate sort of standing at some distance away showing very little concern with his arms folded; perhaps more upset he is losing his night out. Descending down the strip, a fever-pitch buzz

comes over us and there is a humid smell of body odour and vomit as empty bottles are randomly kicked around. We amble down, getting squashed and knocked, almost losing each other in the process, and declining numerous offers from PRs who spend only seconds trying to coax in the flood of potential customers on the strip. As I walk past, I see the same waiter in one bar who has worked there three years running; some things don't change. We get to the bottom, where an ambulance is parked. It is from the private clinic, Médico Galeno, on the seafront. As we look inside, a young girl sits strapped into a seat with her head hanging over her body; she slumps like a dead body while her friend anxiously talks on the phone.

As I look around, the taxi queue is miles long as it is obviously time for the clubs to open. We walk past the prostitutes and the restaurant area and the buzz disappears and only periodic shouts, bottle smashes and screams can be heard. We sit in a take-away restaurant near el huevo which is littered with half-eaten burgers, chips and cans of beer. As we return to the West End, the noise levels start to increase – as if we are approaching a full football stadium. The music hammers out as we walk up and decide to go for a drink in a nearby bar/club place. As we enter, I step onto broken glass all over the dance floor. I put my elbows on the sticky bar to order a drink and shout at the bar woman. On the screens in the bar are images of half-naked women dancing erotically. The music changes and it is one of those songs where the vocal has been modified using vocoder and where the singer isn't really singing but just flashing around their perfect body on the video. As an R&B classic comes on, there is a rush to the dance floor and the grinding starts as people side step on the broken glass. The girl groups keep to themselves while the young men start to circulate to see who will respond to their gyrating hips. As more drunken men enter the bar, they start dancing as if they need to create some impression and to attract attention to their grand entrance. Unfortunately, this seems to do the opposite as, over the course of 20 minutes, the number of young women diminishes. The R&B music continues and the more grimey[1] it gets, the more grimey the men's moves get. It is now after 5 a.m. [Field notes]

A short ten-minute walk can take you to Ibiza Rocks, the famous hotel music venue constructed in 2005 and following around the rocky coastline, past the marina, one passes some greater attempts to insert higher class bars. This continues to the Café Mambo and Café del Mar area where, on most evenings of the summer season, gather hundreds of tourists to watch the sunset. The next section looks specifically at how a particular type of tourism dedicated to the British has evolved over the years in San Antonio.

The commercial evolution of San Antonio

There seem to be three intersecting tourist waves in Ibiza which have been: (1) a slow sprinkling of tourists during the 1940s to 60s which were a cultural and artistic elite followed by hippies respectively; (2) a more pronounced increase with the advent of package tourism and group bookings in the 60s and 70s; bolstered more recently with (3) independent/weekend travellers and larger groups through cheap flights at the turn of the twenty-first century. Here I explain a little about how these movements came about through the testimonies of those who experienced them.

During the late 30s, 40s and 50s, Ibiza was a refuge for German, Italian and Spanish artists and politicians who were escaping the increasing tide of fascism. In interviews with old hotel owners and tourist shop workers, this period, just before the advent of mass tourism as a means of the main threshold of the economy, was reflected upon fondly. In the words of one hotel owner, the early tourists were *'classy'*. Then in the early 60s, hippies and beatniks arrived in San Antonio, learning of a new 'cool' place from the children of U.S. diplomats who had taken holidays there. Latin Americans from Argentina and Uruguay also arrived as refugees and throughout that decade, other Europeans started to retreat to Ibiza's shores for alternative practices of love and liberation. However, as recalled when I went to meet one of the oldest hoteliers in San Antonio, it was also in this period that mass tourism started to develop:

> I climb four floors of the hotel to meet Pablo who rarely gets out these days. The room is long and I glance out beyond the curtains which flap around in the open windows to see a commanding view of San Antonio. Pablo then emerges, shuffling his steps and with him in one hand is a bottle of rosé wine with two glasses. As we start to talk about how San Antonio has changed over the years, he pours me wine and taps his hand on the table to make significant points such as the advent of mass tourism. He says: 'Before the 50s, the British that came were rich and good people. The British that came with these cheap holidays from London after the 60s, and we are talking into the 70s and 80s when they came more in groups, it brought more of a bad name to San

Antonio and this form of tourism continued. Then what we had about ten years ago, maybe more, was the tour operators approaching other areas on the island to build hotels.' When he speaks about the 1980s in particular, his eyebrows frown and he starts to thump the table harder while, at the same time, the old skin under his arm and chin swings around. 'San Antonio started to go elsewhere' he says, meaning that its 'better' tourists migrated at this point. [Field notes]

One significant change seemed to come with the advent of mass tourism in the 1960s. Hoteliers like Pablo were approached by tour operators and asked if they would receive British tourists, and also asked if they would extend their properties – or even rebuild them – to cater for the increasing demand; the premise was to make as much money as possible at the expense of basic maintenance and repairs. When the same operators approached hoteliers in the 70s, asking if they would receive group bookings, this signalled another shift as many of the hotels barely broke even from these contracts and couldn't maintain their hotels to a decent standard when things were broken. Greater numbers of bars and clubs were erected and the West End, as a nightlife hub, started to grow. Little by little, it expanded in the 80s because the local authority was only concerned with making money so more bars and restaurants were built. In turn, the San Antonio Bay started to get populated by larger hotels. Yet there seemed to be no regulation on the number of bar and hotel licences which were issued. The 80s and early 90s also marked another significant change with the advent of the Balearic beat and rave scene which further attracted the interest of British youth to Ibiza and, with this, came problems of drug use, excessive drinking, and general disorder. Although the economy was booming, it was to have a significant impact on the reputation for the resort. Javi, who has lived and worked in San Antonio all his life, said:

> *Javi*: As a consequence the 'better tourists' stopped coming because of the bad name. This was when San Antonio started to get worse, and this was about 20 years ago [beginning of the 1990s]. The blame was also with the local authority because here we have small provinces but each was afraid to back down to higher powers [politicians] which were basically telling them to allow the tourism to expand. It was all about money.

Much of this expansion continued into the 90s and did so with little regulation. However, it was not restricted to only British tourists for

German and Italian tour operators also had a stake in the resort of San Antonio (and a few still do today). Similarly, the emphasis was instead turned on the summer tourist market – because of the larger amounts of money to be earned – and as a result the winter market, made up of primarily mature tourists diminished. This, as one tourist shop owner in the back streets of San Antonio said, pretty much placed the island's economy under more pressure during the summer months to make money to keep it ticking over. Eduardo, who owns a small tourist business, commented:

> *Eduardo*: The problem is we only promote party, drinking and nightlife. The politicians are guilty of this because we have not found another way to market our island. Some years ago, we used to have flights during the winter; a mix of retired Spanish, Germans and some British but they don't come any more because there is no money in it. A plane of 30 people earns nothing which is why they need to be full.

It was in the San Antonio area where most of the tourism started to grow quite quickly towards the end of the 80s and early 90s. Then the global, commercial forces of KFC and Burger King were absent – but this was not to be for long. Quite quickly, it was realised that the British were significant spenders and large amounts of money could be made from their holiday intentions for a 'blowout' (Briggs et al., 2011a). This growth was exacerbated further in the late 1990s with the emergence of 'Cream events' run by British brands which began to market commercial dance music events in the international tourist resorts like Ibiza. This created new demand for these products reflected in the increased availability of all night parties peppered with alcohol, drugs and promotions with inducements for sex and drinks at low prices (Calafat et al., 2010).

There have since been further changes to the nature of the tourist movements in San Antonio, and Ibiza in general – notably the advent of cheap airlines at the turn of the twenty-first century. This meant large numbers of people could visit Ibiza for a couple of days or long weekend, needing only their passport and wallet. Such is the growth of the cheap-airline industry that the small Ibiza airport now services 96 cities across Europe and beyond. Ibiza is therefore no longer a destination exclusively for the elite; thousands of 'others' can now access the Superclubs, the bars, the nightlife, the fun and excess. They can do so for a weekend of wild behaviour as they can on a package holiday; they can stay in hotels for as little as £8 a night or rough it on the beach, or even party through without sleep. Crucially, the tourist does not need a significant amount of money and this is perhaps reflected in recent research which has found that tourists are now taking shorter breaks on the island and spending less when they holiday there (Payeras et al., 2011).

The fact that San Antonio is now largely 'British oriented' is no accident. Many of the German and Italian tour operators pulled out of the area as the reputation of the resort diminished. And while Germans, Italians and Spanish do still visit San Antonio, hotel resorts dedicated to their parties are now generally absent. Eduardo continued:

> *Eduardo*: What has happened over the last 15 years here, and this is also relevant to places like Majorca and the Costa Brava, is that other tourist firms which may offer a more quiet, relaxing break have stopped promoting their holidays in places like San Antonio because it would not mix with the young British and the things they do, the drinking, the drugs, the noise. A tourism of people relaxing or seeing cultural sites is incompatible with one which is promoting party nights ... Scandinavians? Practically disappeared. Germans, practically disappeared. Italians, very few. It is a mono-tourism – young British.

Another restaurant owner described the process as a 'cancer':

> *Tiago*: They [the young British] contaminate the place, no one else wants to come. It's like a cancer, growing and eating – it is poisonous. Each year it all grows because the other form of tourism dies off even more.

So what has happened over the last ten years in the resort is that the tourist diversity has become more homogenous (i.e. fewer Germans, fewer Italians, more British) and the more cultural outlets like traditional tourist shops and local Spanish restaurants have started to turn over ... unless they join the market to compete for money from the British: how do they do that? Open up a bar, club, fast-food place, an off-licence, etc. Gradually, San Antonio has started to lose any alternative way to make money and this is perpetuated further by the reconfiguration of public space to support brand outlets, Superclub shops such as Space and Pacha, global chain restaurants like Burger King and KFC, exclusive beach hangouts (which cost money to enter) and entrepreneurs who open up private hotels such as Ibiza Rocks. And while local Spanish workers moan about this trend, they reluctantly concede it is 'business' and they see little other way out of this economic cul-de-sac. There is a real feeling of defeatism among these people, a feeling of political abandonment.

At the laundrette

> Walking out of the hotel and bearing immediately left, one finds the local laundrette; a curious place piled high with old washing machines and with a suspicious circuit of cables running like snakes across the floor. I talk first to the one [member of staff] on the counter and explain the nature of my research. Most of the business they receive from the Brits is through the Club 18-30 hotel Casita Blanca just down the road; the clothes coming ripped, and stained with vomit, blood, piss, shit and various other bodily fluids. They ask no questions because it is 'business', adding that it is the large groups who use their service. Another man then comes into the picture as he folds a large pair of men's underwear: 'they are getting younger over the last few years', while the busy one says 'They may come for a summer and just stay but can't find work'. They then start to complain about the noise they make and how some tend to aimlessly wander into roads without looking when drunk. The man says 'Some people complain but what can really be done?' 'It is what it is' says the pacey one, shrugs her shoulders and tends to another load. [Field notes]

In a local tourist shop

> After my radio appearance, I receive an email from a local resident who is a member of a commercial group in San Antonio. He invites me to speak to him anonymously in his tourist shop, just off the main drinking strip on a quiet alleyway. I walk into the shop which adjoins a suitably local restaurant and find cultural trinkets for sale; I am far away from sun cream and the *Daily Mail* here. He walks in with a coffee and greets me, peering below his glasses in the process. '*¿No eres periodista?*' (You're not a journalist, are you?) he asks and I reassure him. I sit on a small stool which would otherwise be used to reach the top shelves of his tourist shop – probably only to wipe the dust away rather than retrieve something which was on the verge of being purchased. The main things he has to draw to my attention are the way the British are '*manipulated*' into intoxication; he passionately describes them as '*victims*' of private tourist organisations.

> He says that as a consequence businesses like his collapse because the tourists have no interest in what they have to offer so to survive, people in the local economy more and more have to adapt to tourism which is favoured by the British. [Field notes]

Moreover, this has all been exacerbated by the global economic crisis of 2008 which leaves some hotels and buildings on the island sitting half-complete as developers and investors struggle to get credit from the banks to finish the projects. The crisis has not only decreased local property prices and the number of visitors to the island but is also perpetuated by globalisation of tourism and the increasing competition with other holiday destinations (see Meethan, 2001). The race is on to generate profit and contain tourist spending (Miles, 2010). In particular, the introduction of all-inclusive hotels, as an example of this, has led to an aggressive competition for tourist spending in San Antonio. This astute PR manager who had worked in Ibiza for over a decade said:

> *Oliver:* From 2001 to 2006, the economy was growing, lots of money, they were the best years. Lots of people, it was more people from different countries, not only British but now it is more British. The problem was when they started to include all inclusive because the bars started to lose business so they were looking for ways to tempt people back in so lowered the deals and the quality of alcohol. Because if you are all inclusive, you don't leave the complex because you have free drinks, you don't go out and spend. This changed the way business was done because increasingly more PRs started to appear to persuade people into bars, clubs, restaurants, everywhere. This also meant they reduced the wages of the people working in and for the bars and made them commission only. This made it unattractive and created a kind of lazy workforce ... All inclusive ruined the island. It turns up the pressure for other parts of the economy when this kind of tourism is promoted because no one else wins. People don't go places and spend, they don't see different things and spend. They don't go to restaurants, they don't go to shops so they shut down.

As the campaign for tourist spending increased, the cost of the alcohol was lowered and increasingly deals were introduced on the West End. Moreover, rather than seeking to increase tourist numbers – thereby requiring an expansion of the infrastructure – the political economy is instead looking to attract a new tourist profile which 'spends more money' in the limited time it is there (Payeras et al., 2011). And while the island's economy tries to change its image and clamp down on the problems caused by the tourists, it is constantly reminded that it has got so used to the economic benefits that it is difficult to find another way to withdraw from what currently sustains it. However, as we will see, these attempts to change the image in Ibiza have been largely countered by the aggressive Marketisation of the island in Britain (and online) as a destination to *holiday* as well as *work*, and this is the most advanced phase of tourism which is currently taking place (Chapter 8). One Ibiza DJ said:

> *DJ Fred*: In Ibiza you've seen it when it was underground, its first original peak was when Ibiza was talk of the underground, then it went down a bit because of the drug connection and it's gone through the roof now. This year, along with several other countries which have been seriously promoted and marketing wise, it's gone through the roof – the tourism rate, they're laughing. The credit crunch is bollocks because people are spending money, prices have gotten higher. With Ibiza I really feel it is the hype on the streets; it is the talk on the town, it is everybody wanting to go to that hot destination.

The attraction is not only the music but the club, the aesthetic design of the club, the dancers, the entertainment and the style which is connected to the clubs, the 'names' and the branding (Meethan, 2001; Miles, 2010). Because of this, we are not just seeing 'hardcore house' music lovers coming to San Antonio, and different parts of Ibiza, but swathes of different tourist groups.

The British tourists

These days, while San Antonio is seen as a holiday destination for young male and female groups, it is now accessible to the masses. While some have holidayed in other destinations before coming to Ibiza, others are now spending their first group holiday there (Chapter 6) largely because of the aggressive marketisation of the island to boost flailing tourist numbers. These groups come either through cheap Club 18-30 packaged holidays or independently book through cheap airlines and local hotels. Generally these sets of tourists have only one thing in mind: to get 'on it' and spend their nights mixing West End experiences with access to the Superclubs. They may also go to some of the all-day beach clubs to continue the party or do so in their hotel rooms.

They proliferate throughout the season with the more serious clubbers coming for the opening and closing parties in June and September respectively. Increasingly, as with other European destinations, stag and hen parties play a role in this tourist landscape (Thurnell-Read, 2011). They range from young to more mature groups of men and women. Typically, they may be up to these sorts of activities on the West End:

> To my left, a circus of middle-aged men stumble out of the bar where we drink; each dressed in some sort of fancy-dress costume. As they come out, they all point at a half-naked man who sits with his face on his arms. They push him into a sitting position and he still doesn't seem to recover; his mouth gapes open as if he is fast asleep. One of the group reaches into his pocket and pulls out a red marker pen and writes something on his forehead. The poor bald man just sits there as the giggles turn to outright laughter; even the waitress joins in. I go over to see what they have written and as I arrive, the waitress adds something on his chest: 'cockless' she writes while scribbled in capital letters on his forehead is 'SHITCUNT'. The group are in hysterics as it transpires it is his stag party and he has lost his costume. Shortly after, as the laughing continues, he comes to and raises a docile smile as if to suggest he is 'the cool guy'. He gets up and high fives his friends who refrain from telling him about the offensive graffiti they have applied to his face. [Field notes]

The tourist demography has thus shifted substantially over the years and this has significant implications for how the resort of San Antonio now has to function (Chapter 8) and the deviant and risk behaviours which ensue (Chapter 9).

Conclusion

This chapter shows that the last 30 years have seen an expansion in tourism across the island and these days the tourists who visit Ibiza do not only go for the 'music' nor go because of its 'hippy roots' because a rampant commercialisation has enveloped the iconic cultural attributes for which Ibiza is famous (Boorstin, 1992). This commercialisation has attracted private investors as it has illegal networks (Chapter 8) and has meant that the economy is almost exclusively reliant on a party-oriented, summer-month-only tourism which also buoys the economically stagnant winter months. During these summer months, the resorts which cater for the nightlife and unlimited excess become problematic

(San Antonio and to some degree Platja d'en Bossa) because of the volume of predominantly British tourists who descend on these small areas with very similar intentions to get wasted. One major problem this creates is that the economy becomes almost irreversibly reliant on this form of tourism and the level of spending which accompanies it. This is at a time when Ibiza is in competition with other holiday destinations (Meethan, 2001), has seen a reduction in tourist numbers, and withdrawals made by the Italian and German tourism operators in San Antonio. So as the heterogeneous balance recedes, a more homogenous one evolves around the values and interests of the primary tourist group – the British tourists – and does so in the form of bars, clubs, off-licences, beach clubs, strip clubs, etc. (Billig, 1995). These changes create large strains in the political economy and prompt complaints from residents and local businesses about problems the British bring but their voices are frequently ignored because the island operates under free-market capitalism where profit comes first (Chapter 3). As I have said in Chapter 1, what these young working-class Brits do when they travel abroad is as much about what they do on holiday as it is what they do at home. The next chapter explores how these areas of home life and leisure pursuits have become integral parts of their *habitus* and this, I argue, lays some of the foundations for their behaviours abroad.

5
Goin' Ibiza: Home Lives and the Holiday Hype

Dan:	What is it that you all do in the UK?
Jay:	Building.
Nathan:	I just come out of the army so I am like proper stressed. I come out of the army, see people get blown up, so I have a lot of stress.
Jay:	That's why we've got to have fun, mate. So if he's not having fun, I'm not having fun. And we are on holiday. If I've got £100 and they have nothing, I'm gonna split it.
Dan:	Right. [To Paulie] What do you do at home?
Paulie:	Construction work.
Marky:	Yeah, me too.
Jay:	We all do the same sort of thing.
Paulie:	We're all riffraff [of low social calibre].

They say their work is 'boring'. It then transpires that all have criminal records and have spent a significant period of their youth in the weekend local NTE, drinking heavily, and taking and dealing drugs, and getting into fights. Later in the interview:

Jay:	Yeah, we all used to smash it on drugs. When I was 18, I was on it non-stop. Proper on it. For a whole weekend, it would keep you alive. Say from Friday to Sunday night, it would be non-stop and you could drink more.
Nathan:	It's true.
Jay:	If you want the real truth, this is what we do. I wouldn't lie to you. Me, him [Nathan] and my mate in three hours, we finished 21 grams of coke.
Dan:	How much was it worth?

Jay: Easily, that was strong shit, easily a grand's worth.
Paulie: I went to jail for selling it. That's why I don't touch them. I come out two years ago and won't touch them because I lost so much through drugs. It fucks you up, mate.

Introduction

The experiences the Southside Crew described here are similar to numerous others in my sample of working-class Brits who go to San Antonio. Many reflect on growing up and being familiar from an early age with drinking, drugs, drug-taking and, in some cases, drug dealing. Those that have work tend to describe it as tedious and mundane while others survive more by temporary and uncertain means (Standing, 2011), perhaps augmented with illicit activities such as crime (Hall et al., 2008). These precarious positions are complemented by an attraction to getting drunk and/or 'living for the weekend', perhaps getting arrested, and/or engaging in deviance and risky behaviours. What it is therefore important to acknowledge then is that these practices are already embedded in their cultural tastes and life attitudes – their *habitus* – and this, to some degree, influences what they do on holiday (Chapters 6 and 7). However, I want to show in this chapter how these elements of their *habitus* have, over time, been moulded – and are increasingly shaped by – the delegitimisation of work and labour, and instead complemented by an aggressive commercialisation of their leisure time which reflects a shift towards a consumer society (Chapter 3). I want to suggest it is this socio- and culturo-structural framework which blinds them with a thin film of ideological fantasy about how they should enjoy this time – that is getting drunk and/or taking drugs at weekends, and engaging in deviance and risk behaviours in places like Ibiza (Chapter 9).

Daily routines and responsibilities: Banality, familiarity and the quest for transgression

For the majority of my sample, home life – generally consisting of either 9–5 jobs, looking for work, time between studying, partners and family responsibilities – is constructed as boring and mundane. For the majority of employed men, work is generally in the manual labour and service industry such as construction, plumbing electricity, retail, etc. with a few having studied for degrees and working in cities in junior management positions. For the young women, work is mostly office/clerical, in the local NTE at home, retail, or various positions in the beauty industry. Those between work or looking for work tended to be either studying at university or occupying quite precarious positions between 'cash-in-hand' jobs and life in the illegitimate economy (mostly topping their wages up through drug dealing). As we have seen, this was certainly the case for the Southside Crew. At home, they confess to getting into fights and all four have been in trouble with the law; Paulie has served two years in prison for cocaine dealing and Marky three years for grievous bodily harm (GBH). Jay *'had a gun'* and 'used to deal cocaine in kilos ... it is a different world. Before I knew it, I was dealing millions.' When Nathan has been *'desperate'*, he intermittently grows cannabis at home and makes up to £5,000 at a time.

There seemed to be a general consensus in my sample that the role of work, if they had any, was to sustain leisure pursuits and that every opportunity in their youth had to be seized to celebrate the moment before either responsibility and/or old age started to interfere. Most prominent in responses was a commitment to drinking and/or drug-taking at the weekends and that these pursuits countered the weekly problems of banal work or looking for it. Many seem to think that if they were not doing these things, that this would mean they would otherwise be partying. Marcus, who works for a gas company, said he confined drinking to weekends because he has work to contend with: 'If I didn't have work, I would drink all the time.' Equally, some of those with both families and jobs also felt this way. Here two working, single-mother students, who had been to various other holiday destinations and to Ibiza only once in 2010, reflect on this dichotomy:

Tina: People have responsibilities, like work. Work is the major thing, like work is boring, work is mundane. You think 'I can't go drinking tonight' because if you do then you have to worry about getting up for work tomorrow, being with people, like a lot of people's jobs you have to deal with other people. You can't do that [with a hangover].
Sharon: Especially when you really want to tell them to fuck off.

There was some evidence, though not much, that being in these positions was frustrating and one of the only ways to accept them was to indulge in weekend excess. There exists a pressure among the cohort to therefore make the most of their leisure time; to seize its preciousness as it, as they construct it, represents some form of freedom from what they say is a restricting and boring home life. It's a similar story for Liam and Graham who are in their mid-20s and have been to Ibiza three times in the last year; twice for the closing parties and once for a holiday between. They say they go to Ibiza because of the music, because it's mellow and chilled out. Although they exhibit exactly the same behaviours as they did in Zante when they were younger (Chapter 6), they describe work as *'fucking shit'* and *'boring'*. At home, drinking, drug-taking and deviant and risk behaviour is governed, to some extent for some, by work patterns – heavy drinking at weekends is followed by some period of self-reflection, remorse and regret but over the course of the working/non-working week, for many, an internal pressure bolstered by social expectations to go out starts to build again. Most don't seem to recognise this pattern – like Jane and her party of four friends whom I interviewed on San Antonio beach:

Jane: Like you drink on the weekend and then you regret it and you don't drink for a while but then you end up doing it again – like the week after or something.

While a few, like Graham, were more reflective about what was going on:

Graham: Ninety per cent of the British public have been tricked into working all week for the weekend, going out and get completely hammered off their face on Friday and Saturday night, sleep on Sunday and then go back to work on Monday and moan about their own life. They do exactly the same thing the week after. That is not how it should be.

Graham is clearly intelligent here in that he identifies something quite significant about how people like him are subliminally governed by the social structure to part with their money and other resources for the sake of propping up the various economic industries (drink, music, club, etc.). Yet, he knows little else than to suggest that it shouldn't be like that. Other recent changes to the ways in which this cohort celebrate and engage in excessive consumption are often at festivals, stag parties and, more relevantly through the advent of the holiday. These other 'new' means of hedonism also provide extra impetus outside the institutionalised regularity of weekend excess (Presdee, 2000); they are something else to have on the horizon while home tedium plays out. However, there are more complex socio- and culturo-structural processes and subjective associations at play, and in the sections which follow in this chapter, I want to indicate that it is these other elements which, in part, lay the foundations for the behaviours which this group of British youth exhibit when they holiday abroad. Firstly, I want to suggest that it is the way in which these practices are socially embedded in the lives of British working-class youth.

Socialising intoxication: Some case studies of how deviance and risk become embedded in home life

I have stated earlier that what this group of British working class do while they are away is as much about how/why what they do at *home* as how/why what they do when they go *abroad*. We have seen in recent years how the NTE, and the marketing surrounding it, has been cleverly bent towards directing young people into these practices, and how, in turn, this has influenced attitudes to intoxication (Chapter 3). Coupling this marketisation of intoxication with working lives which, for most, are considered to be mundane and boring, or to some degree precarious as others are in and out of work/study, makes for a recipe of wellbeing generation, social kudos and, as a consequence, subjective identity construction around leisure, 'play', deviance and risk. These three brief case studies augment this supposition.

The 'beachgirls': Drinking stories and creating 'memories'

It is just after 4.30 p.m. in San Antonio, and I decide that a large group of girls on the beachfront would be ideal to approach. A few are unemployed but most work in the beauty industry while some are studying. A few are also friends of friends and don't really know each other. As a group they often go out on the weekends back home in Kent in south-east England. The 'beachgirls' are all aged 21 and most went to the same school. Most Friday and Saturday nights start off with a *'pre-drink'*, often made up of a bottle of wine each or a litre bottle of vodka between them. Although they say their mums think it's a lot, they say 'we think its normal'. They get 'really drunk to start with' to save money. This starts about 8.30 p.m. until about 11 p.m. When they get to the clubs they are *'hammered'* and spend then around £30–40 which includes money for *'cabs, kebab or burger'*. They say the aim of these nights are to 'to have fun, have a dance, create memories', one adding that 'When you are older, you can't do it'. Boys, it seems, are not really on the agenda. What often happens is that they will take photos of each other and their drunken escapades and tag each other on Facebook. One reflects on a recent night out when she fell off her friend's shoulder and broke her tooth while another recalls how she was *'so drunk'* on one night out recently that the bouncers threw her out of the club. For them, it seems funny all the same.

The 'unpredictables': Powering through the weekend

The 'unpredictables' are a group of four young men aged between 19 and 20 who live in London and, aside from some student nights on Wednesdays because of the *'cheap drinks'*, confine drinking, and the rest of it, to weekends. One is attending university while the other three dropped out and instead have part-time jobs in retail stores. Spending between £30 and £60 on a night out, they all seem to display a sense of *'powering through'* their nights out. Kev said: 'if I'm out Friday, do an all-nighter and work through. If I go to sleep then I won't wake up'. Aaron, who starts work at 7 a.m. on Saturdays, does the same on Fridays except he often goes home after work, 'sleep after work for a couple of hours and then go out Saturday night'. They agree it is *'pointless'* to go out sober. They spend and drink more on paydays, and are more likely to treat themselves to a *'sniff'* (cocaine) as well, one adding that 'the more you spend, the better it is'. They don't plan where they go as sometimes nightclub bouncers won't let them in because of their reputation. They think a good night out consists of how much they drink and who they *'end up with'* at the end of the night (kiss or have sex with) or even if they find themselves *'looking for madness'* (a fight). Either way, says one, 'it is something to talk about the next day.'

The seasoned weekenders: More condensed partying

Gav and Fred are two young men in their late 20s. Much of what they did as teenagers was not too dissimilar to the 'beachgirls' and the 'unpredictables'. As we sit drinking some pints in a south London pub, they reflect that from the ages of 16–19, at college, they both used to be out drinking and taking drugs most nights of the week which included *'three-day benders'* at weekends. They remembered that they 'always started with beer' – a *'warm-up'* but that it quickly moved on to shots because they 'used to get bloated from the beer'. They did this because the 'booze was cheap' and wanted to 'have fun with mates' but that when they reached their early 20s, they realised this level of partying could not go on forever as they got jobs in the construction industry. Midweek drinking and partying faded out and was then reserved for weekends. They regret spending so much money but then counter it with 'Well, they were good times'. Now they go to the pub as early as possible on Fridays and all day Saturday for the 'blowout period' and smoke weed intermittently between these sessions, adding that the holiday is the *'catch-up period'* in which they 'make up for lost time'.

These short examples show several important aspects of the cultural practices which the people in this book start to assimilate as they grow up in the UK. The first thing to say is that there is a general attribution, and acceptance of in some cases, that engaging in risk and deviance is normal and a 'fun' thing to do. For many, it gives them something to say – a story to tell – one which can be recounted and used throughout the working/non-working week (and beyond) as a measure of the enjoyment they had. These stories often dominate pub and office time, as well as virtual space, which keeps alive their significance. However, as the examples collectively concur, there is some morbid determination to be in these states.

In addition, while some patterns of excessive consumption change according to disposable income and, perhaps more longitudinally when the responsibilities which accompany the transition to adulthood (work, family) accrue, it doesn't necessarily mean they reduce or disappear; instead these ambitions to engage in excessive consumption are either condensed and/or reserved for particular times – of which the holiday is one – and this is in part why we see varied British holiday demography in Ibiza (Chapter 6). Importantly, however, this is when people tell themselves they deserve a 'good piss-up' or they need to 'get on it' because otherwise they feel they are missing out; they aren't/can't seize the moment like everyone else (Chapter 3). They need to do this to satisfy their individual desires but also create envy from their participation in them. The vehicular forum which makes this possible is often through social media forums like Facebook and has few boundaries when it comes to attracting people into a night out (or even out of nightlife retirement!). In the context of Ibiza, for example, it looks a little like this. Having graduated through ticket selling and 'booze cruise' stewardship in one season in Ibiza, Tim posted this on the eve of obtaining a job as a PR:

Tim's status:	This is when my dreams come true (13 likes)
Wayne:	Smash it!!
Steven:	You're a very lucky bastard, I wish I was there to witness this!
Lewis:	Would love to be there with ya bro you lucky bastard!! Av fun mate ☺
Steven:	Bet its going to go off.
Deano:	Just seen this … what's goin' on? What's Tim doin'?

There are no further qualifications from Tim because this is part of the suspense which maintains a social interest in what he is doing as well as the collective envy which comes with it. It is a social envy of intoxication, transgression, living life to the fullest which seems to be driving others to feel they need a piece of the action (Chapter 6). In fact, during the summer of 2012, and perhaps by no surprise, several of his friends visited and stayed with him in Ibiza (see Chapter 10 for what happened). However, the backdrop of this cannot only be that these norms and values of this group just *exist* and are shaped exclusively by the NTE industry because there is, I want to add, a further layer of ideological influence which helps ensure that leisure, excess and intoxication trumps and therefore devalues education, training and work as a means for identity construction: popular culture such as TV, the media and the internet. Collectively, these cultural modes promote ideas of intoxication, party, playtime and for the people in my book, it is these ideas to do these things which surface when leisure time becomes available.

The only way is (unfortunately) *Essex* ... unless it's *Made in Chelsea*: The increasing emphasis of playful forms of leisure in popular culture

Pamela: You know *The Only Way Is Essex* [TOWIE]?
Dan: Unfortunately, yes.
Pamela: It's like that on holiday. Well not necessarily in all Ibiza but perhaps San Antonio. Everyone has five-inch wedges on, bikini and make-up, you know, TOWIE, boob jobs, fake tans, slim.

Here, Pamela describes how the person aesthetics of some popular docu-reality programmes led by B or even C-list celebrities – such as *TOWIE* and *Made in Chelsea* – are reproduced by the masses in San Antonio. Programmes such as these, as well as *The Bachelor, Take Me Out, X Factor* and *Britain's Got Talent* to name a few, heavily feature in the UK and are adamantly followed by millions of people. While different in their orientation, in one way or another, these programmes celebrate and promote the 'good life' which is about:

- Having money without visibly doing anything to earn it;
- Spending that money willy-nilly as if it was magically replaced by more money which can also be spent willy-nilly;
- Material life as mode of *being-in-the-world*;
- Projecting social envy as a means of social status;

- Engaging in excess, deviant and risk behaviours;
- A narcissistic sense of self as a means for 'feeling good'.

In programmes like *TOWIE* (catering for and featuring working-class young people) and *Made in Chelsea* (catering for and featuring upper middle to upper class young people), the cast are caricatures of young people in their early 20s who have amazingly perfect bodies, groomed eyebrows, really white shiny teeth, eyelash extensions, and long nails. They show off the latest designer labels *a la mode*, live in nice, large houses with all the latest fixtures and fittings and have youthful-looking families and friends. In *TOWIE*, however, there is no essence of work or responsibility or any authentic substance of daily life which is closely familiar to what most people supposedly in this bracket would experience in real life. None of these people live on shitty estates, are single parents, have criminal records, and, as a result, struggle for income and self-respect in low-paid, low-grade jobs. Instead, the caricatures are followed around by a wobbly camera as they swan around in nail bars, beauty shops, gyms and nightclubs, where they basically just talk about sex, drinking and relationships. This is not exclusive to the everyday locations in the UK either. In fact, the main star of *TOWIE*, Mark Wright, DJ'd in Lineker's Bar in San Antonio in 2011, while two *TOWIE* stars, 'Arg and Dials' and one from the *X Factor* runner-up pop band JLS, partied in Ibiza over the summer of 2012.

The Bachelor and *Take Me Out* offer hegemonic determinants of what young people *should* look for in a relationship. In *The Bachelor*, numerous young women – mostly with the kind of physical and style attributes I described in the previous paragraph – have to awkwardly vie for the celebrity man who is not really interested in any of them but has to pay them a series of meaningless compliments for the duration of the programme. They seem to do this by using their feminine charms to woo him – after all that's all he is interested in. In the process, each week, a woman is kicked off because others are more preferred by the celebrity and the bitchy politics are eagerly but grossly documented by the production team. Similarly, in *Take Me Out*, numerous women – again with the same sort of exaggerated attributes – have to compete, mostly by using their plastic images and predictable one-liners, against others in competition for 'one lucky man'. However, aside from using their visual aesthetics, the discussions and compliments often revolve around 'how good the other looks', 'how nice his car is' or whether the 'style of clothes' are in sync with each other's taste. In both these programmes, a false sense of relationship and mutual relations are provided around materialism, and the women are generally treated as 'objects' or 'trophies' from which the man must select (see Chapter 8). Again, it is no accident that the 2012 series of *The Bachelor* was filmed in various exotic beach locations including France, Italy and Bermuda. Equally, *Take Me Out* has a holiday theme to it as the successful couple/pair have their 'date' on the Spanish island of Tenerife where, more often than not, they find out that they made poor selections as conversations hit an inevitable dead end over dinner.

X Factor and *Britain's Got Talent* offer the chance to any regular person – frequently working class – to 'live the dream' and become a singing/performance star. Entrants go through a rigorous selection process which is measured as much on their singing/talent capacity as it is on 'the way they look' – if they look like a star and have all the kit – and the broken story attached to their motivation for wanting to 'live the dream'. As many of the competitors know, winning and even getting through to the live sing-offs guarantees a short-lived lifestyle of sex, drugs and rock and roll as well as a limited shelf life in the public consciousness. If they pass the first few rounds, they are then flown to the judges' mansions which are located in exotic destinations around the world where they, as well as the audiences, can catch a glimpse of what life would be like in this material heaven. When there is little hope for life in the margins, at least there is hope that there is a way out for everyone who has 'hidden talent' and that anyone can realise their potential just by entering these TV talent programmes.

Over the last ten years, these programmes, and numerous spin-offs of them, have increased in frequency. This has been assisted by the introduction of Freeview television which now offers British viewers over 100 different channels as well as the increasing monopoly, and consequent popularity, of Rupert Murdoch's Sky TV. In essence, it is these programmes which dominate prime time viewing as well as having daily spin-off chat shows and repeats on other channels that are part of the same network. The lives of these people on these programmes, the DJs, and the celebrities are further idolised in newspapers, magazines for both men and women, as well as on the internet websites and blogs and radio. For example, numerous radio stations like Heart, Capital FM and Radio 1 are all guilty of promoting the frivolous lives of these celebrities. They also propagate an ideology that the 'working week is boring' and that the 'weekend is the release', often encouraging the public to phone in about how they skilfully pulled a sickie on Monday morning after their three-day benders at the weekend. The DJs play music tracks which try to liven up the Monday morning blues, then, throughout the week, start to hype up the weekend nights out and have endless holiday competitions for places like Ibiza. In short, there is little escape from celebrity lives, what they do and how they look. It is these cultural modes that collectively guide the interests, desires and fashion styles of the people in my book and this is as evident on any Friday or Saturday night in UK town centres (see Chapter 7 for how this plays out in Ibiza). Indeed, as Blackman (2011) suggests, it is this celebrity culture which guides attitudes, values and behaviour in the context of intoxication practices. No wonder people are queuing up to 'live the dream' in Ibiza or some far-away distant land where work is play and play is work. The same cultural mediums also have something more direct to imply about British behaviours abroad.

Populist media portrayals of 'Brits abroad'

Among the hundreds of news articles about the stampede for the sun on the Costa del Sol and other sumptuous European destinations, a significant number of UK newspapers report on the deviance and risk behaviours caused by British tourists.[1] Dating back to the 1990s, most report on a mix of social problems such as consular problems (tourists losing passports, travelling without insurance, requiring money), accidents (self-injury, death), violence (between each other, against locals and in bars/clubs), and drug use and drug dealing (in places like Cyprus, Majorca and Ibiza). Yet all seems quiet on the global corporations and tourist companies which capitalise and endorse spending, excess, deviance and risk (Chapters 6 and 8).

Despite this negative reporting of the problems young British holidaymakers cause, the same newspapers also promote the same holiday destinations as idyllic retreats and places to enjoy the 'best beaches' and 'best clubs'. For example, writing on the Balearics in June 2000, early in the holiday season, Charlotte Adams reports on the beautiful beaches, cultural heritage and archaeological sites, even recommending Pacha as the 'coolest club';[2] while only a few weeks later, an article in the same newspaper bemoans changes to the way British youth drink and the extent to which they get intoxicated on holiday in the coastal resorts and islands of Spain.[3] The same occurs the following year when *The Guardian*'s Nick Green writes persuasively about the amazing 'clubbing experiences' to be had by young British people across the Mediterranean (even listing them in the process);[4] yet, just over two months later, his colleagues Audrey Gillan and Mary O'Hara reported on the Britons who are arrested for drug dealing in the very same nightclubs in the very same destinations.⁵ It is this tension which exists in most of *The Guardian*'s portrayals of holiday destinations for British youth.

The Sun newspaper's portrayal of holiday destinations such as Ibiza also offers a similar contradiction: on one hand, the depictions report on the problems attributed to British tourists abroad (mostly to do with drunken behaviour, drug dealing, fatal accidents and deaths) but, at the same time, cannot help but popularise, and almost canonise, the same destinations as fashionable places to *be*, largely thanks to the celebrities who go there and behave in the same way as the tourists. For example, in the summer of 2012, a 19-year-old British tourist was found dead in a hotel swimming pool in Ibiza⁶ while, only a day later, another Brit died after attempting to jump off his Ibiza hotel terrace into a lake after a drinking session.⁷ Yet, in the same year, Tulisa Contostavlos, the UK pop star and presenter of *X Factor*, was reported to have stayed in the San Antonio resort of Ibiza Rocks, appearing mostly in tabloid news media for her drinking sessions and partying while her fellow N-Dubz⁸ colleague was quizzed by police after arguing with hotel staff and allegedly having sex with five women in his hotel room.⁹ This was also epitomised by the Kaiser Chiefs' lead singer who, when playing at Ibiza Rocks the same summer, jumped off the balcony into the swimming pool – much to the anger of security but to the jubilant pleasure of the young British crowd staying in the hotel. Of course it is difficult to suggest that young working-class British tourists are adopting these behaviours *because* of their celebrity idols but the fact that the latter group are up to the same thing speaks volumes about what behaviours are typically expected to occur when on holiday and are thereafter presented in the media.

Similar behaviours have been/are also popularised in TV and film. For example, *Ibiza Uncovered*, filmed by Sky TV from 1997 to 1999, sought to 'uncover the island which offers sun, sex, fun and much more'. The 'documentary' claimed to 'show the extraordinary ups and downs on the island designed to take things to the limits'.¹⁰ The series followed stag parties, gay men, couples, singles out for some seasonal fun, families, club reps and even older, more mature populations. However, a major feature of the series was to focus on the massive sacrifices young Brits made – as in give up work in the UK – to go out and 'live the dream' by working a season in Ibiza. Similarly, the ITV series *Club Reps*, which televised from 2001 to 2004 across UK networks, popularised the experience of British youth working in a holiday resort. Filmed principally in Faliraki, Rhodes in Greece and Gran Canaria, Spain, the main features revolved around the difficulty the Club 18-30 managers had in regulating the behaviours of their play-hard workforce, while at the same time, disentangling them from drinking and sleeping with each other and the tourists for whom they were responsible. In doing so, *Club Reps* acted as the almost blatant advertisement to other like-minded young British people to drop what they had (if they had anything to hold on to) and go out and 'live the dream', work in resorts where their work was play and their play was work (Guerrier and Adib, 2003).

Reality TV has also assisted in constructing ideologies of attitudes to the holiday and relationships on holiday. *The Villa*, which aired from 1999 to 2003, popularised the idea of matchmaking in the holiday context. Filmed on the Costa del Sol in Spain, the reality TV programme would follow four young men and young women, encourage them to go out drinking and clubbing and follow their movements to see if 'romance' in the form of kissing, groping and sex evolved from proceedings. Similarly, the cameras in BBC3's *Sun, Sex and Suspicious Parents*, which only recently started airing in 2011, follow groups of young British people on what, for most, is their first holiday abroad. Filmed in various popular British holiday resorts around the Mediterranean including Magaluf in Majorca, Kavos in Corfu, Ayia Napa in Cyprus, San Antonio in Ibiza, Malia in Crete and Laganas in Zante, it shows how young and often inexperienced Brits experiment with excess and relationships on their first lads'/girls' holiday. Little do the young Brits know, however, that their parents are watching them drink excessively all day/night, vomit on themselves, and fondle and sleep with members of the opposite sex. Not that it matters as most of the parents – some of whom had been on these sorts of package holidays when younger – seem only mildly shocked about the behaviours, and a few even end up cheering them on as they watch the footage: '*Go on my son!*' says one proud father.

The reality TV industry has also been relentless in fuelling curiosity/interest about the holiday resort. Filmed in 2001 in Kavos, on the Greek island of Corfu, *Bar Wars* was billed as the reality TV hybrid between *Ibiza Uncovered* and *Survivor*.[11] The series' concept was a competition between two bars which had to draw business in to survive in the popular nightlife holiday resort. More recent popularisations have come in the form of the working-class parody of Brits abroad, in *Benidorm*. The series, which started in 2007, follows the same types of holidaymakers in the same all-inclusive hotel in the same resort, year after year. The film industry has been just as guilty in their celebration of deviance and risk behaviours of Brits abroad. Drawing on references to drug-taking, excessive drinking and promiscuous sex, *Kevin and Perry Go Large* (2000) is the puss-in-boots story of two teenagers who drop everything in the UK to seek the high life of infinite sex and DJdom in Ibiza. More recently, *The Inbetweeners* (2011) follows the exploits of four recent high school graduates in Malia. The film is loaded with deviance and risk-taking such as skinny-dipping, having sex with mature women, all-day drinking boat parties, male strippers performing autofellatio, oral sex, drug-taking, fleeting sexual encounters and hospital admissions.

Collectively, these depictions of British tourist behaviours abroad function in a number of ways: firstly, they strategically market different Mediterranean resorts for British youth; secondly, they act as a blatant advertisement to young working-class British people who could potentially be in that periphery category (unemployed, manual work which is monotonous, uncertain futures) to either save all their money to spend on a holiday in places like Ibiza or even put it on credit – after all, as I have shown, a life of leisure as consumers is what they have been socially conditioned to seek (Chapter 3). Lastly, they project a series of behaviours which are expected and commonplace on holiday, thereby reinforcing the normality of their occurrence. We therefore should not be surprised that, when it comes to the advent of the holiday, there is a fever-pitch excitement because this is the chance to do all these things, counter home existence, seize the moment and enjoy the 'good life'.

The 'holiday hype'

I have already shown how the media and other social institutions act to advertise not only the holiday as social occasion for excessive consumption and deviant and risk behaviours but also as a destination where a kind of unlimited permissiveness exists. It is this socio-cultural framework which has produced the kind of group discourses which surround the 'holiday hype' and this helps to comfortably relieve the awkward daily tedium of work – or the precarious attempts to get it (Chapter 3). When I say 'holiday hype', I mean the way in which young British people prepare for (shopping, gym, beauty treatment), 'talk up' the holiday and socialise it into their everyday experience: where they may go this summer, with whom, what's new, what might happen. With the advent of social media and the advances in the technological era, these discourses increasingly take place in forums such as Twitter, Facebook, through BlackBerry Messenger as well as face-to-face meetings down the pub/bar/club. However, the virtual and the accessible nature of these exchanges make it convenient to deviate from a frustrating or boring day at/out of work or to make a comment and bring to the surface the excitement of the weekend drinking session, the upcoming festival or, indeed, the impending holiday. We briefly saw earlier in the chapter how this might work when Tim posted his status on Facebook, but here is another example which typifies many in this cohort.

On Facebook

I met Becky in Ibiza in July 2012 on a boat party in Ibiza. She is in her mid-20s and works as a receptionist in the Midlands area. I kept in contact with her and followed her status updates with interest over the summer of 2012.[12] One typical Friday evening on Becky's Facebook page:

> *A Friday in early May: Becky posts on her wall post a picture saying*: 'Keep calm, it's Friday. GET DRUNK' (1 like)
>
> *Becky*: Waaaaaahahaaaaa! I like this, it's nice. X
> *Leanne*: Ha ha, thought you'd appreciate it. I've just stocked up the ice cubes, ha ha. Vino [wine] needed, last bank holiday for meeeeee xxxxx
> *Becky*: I'm on the regime over here, I've got us a bottle of sparkles. Immense fun is to be had xxx
> *Leanne*: Animal style ☺ ha ha missed your crazy face!!!!! Let the games begin!! Xxx

It is then in June that the build-up starts to Ibiza:

> *A Monday in mid-June status update*: 'Ibiza this time next week' (8 likes)

Nothing is posted in this time. Then:

> *A week later a status update*: 'Ibiza on Friday. Yes, that's nice ☺' (9 likes)

I meet Becky with her friend Jackie in Ibiza early in July. When I start talking to her, it turns out she spent her last €50 of the holiday money on the boat trip but *'has her card'* for the last week of the holiday (Chapter 7). She and her friend flirt and talk with some lads from Birmingham in the heat as we await to depart. In their exchanges, she reveals how she saw a *'massive fight'* in which *'no one did anything* [to stop it]' and there was *'no police'* (Chapter 8). During the trip, they get progressively drunk and before they depart for the West End, we exchange Facebook details. There are no posts or status updates for a few weeks. There then appears:

A Friday in the third week of July: Leanne posts on Becky's wall a picture of a drunk man in a bin. The title says: 'Can't wait to be ashamed of what I do this weekend' (1 like)

Becky: ha ha ha!!! Basic truth! I've got a little treat for us to consume tonight ☺
Leanne: Hmmm, I'm intrigued?! Xxxx

Three days later, to counter the home blues (Chapter 10):

Status update: 'Ibiza round 2 booked. Sweet one.' (4 likes)

Hayley: When??????
Mel [AKA 'MJ']:   In Sept for a week ☺ xxxxx
Hayley: niceeeeeeeee xxx
Jackie: Just uploading pics from round 1 Becky, will be with you shortly. X
Becky: Thanks! Finally I can change the profile pic with a beaut Ibiza one! Been hearing about u I have u shlagggg xx
Jackie: Oh my Beck, you are going back to Ibiza and you are a savage! I need some entertainment in my life! This reality ain't goin good! Just looking forward to basic creamfields snatch now man! X
Dan [AKA me the author]: round 2? As if the sucker punches from round 1 weren't enough ... maybe you were just sparring.
Becky: Ha ha! It was too good! Sparring is all it was ☺

The 33 photos which are then uploaded from 'round 1' include 'predrinking' on the hotel balcony, various ones by the beach and from the boat party, Becky posing in her bikini holding a small sausage as if it was a penis, Jackie with a T-shirt on which says 'Keep true and party on', drunken shots from hotel-room parties with a group of eight lads, and a few with post-night out kebabs. But it isn't long before the posts revert back to weekends and the 'holiday hype':

> *One Saturday in the first week of August, a wall post from Leanne*: 'Six weeks tonyt wil be out n bout the island of dreams waaaaaaaaa aw gutted I am havin stay in. I am again in bed. Let's go galavant tomorrow xxx (1 like)

Then between further weekend party posts, and status updates, it then transpires that more of the 'girlfriends' have joined the 'round 2' Ibiza outing:

> *One Thursday in the third week of August, a wall post from Nat*: 'I actually want to squeal with excitement over Ibiza … 3 weeks tomorrow with Kerrie, Leanne, Charlotte, Suzanne, Mel and Becky ☺' (14 likes)
>
> *Sammi*: Jel [jealous], have a gudden! X
> *Helen*: Boo I should be going as well lol x
> *Leanne*: Extremely excited!!!!
> Nat: You coming now Suzanne? Xxx
> Suzanne: Yes its just I hadn't took chloe away this year [daughter] but she will get a holiday I hope when we're back! X

After a few seemingly frustrating and cryptic status updates which consist of one word, it's back to the weekend madness:

> *Friday end of August, a status update*: 'Giddy' (1 like)
>
> *Leanne*: Why, why why?!?!?! Tell meeeeee!!! Xxx
> *Becky*: Because its nearly the weekend! I LOVE weekendness! Xxxx
>
> *The next day, a status update*: 'Always so parched at the weekend' (5 likes)
>
> *Lewis*: I understand, hello night out number 2
> *Janet*: Me too Beckyyyyyy – its only bloody wine o'clock! Yay!
> *Becky*: ha ha! Weekends are thirsty work!

And the holiday hype continues as the 'round 2' gets closer:

> *The first Saturday in September, a wall post from Mel*: 'Ohh getting closer and closer what u doin to nyt?'

Becky: Naff all, not feeling owt right now. Cany wait for ibiza like, might get my case out inabit xx

Mel: nah i not went twn las nyt, ruf to day and save dosh init. hav u done insurance yet. Nails n hair cut wed book taxi tomorrow packin in process nearly ready to rol baby xx

Becky: Not done insurance yet, will do one night in week, I'm gettin nails done weds lunch, need get few bits n bang it all in case n ready to partayyyyyyyyyy :) allll day and allll night xx

Mel: ye got me oil after sun n deodorant n stuff yest from body care takin shampoo n that got here just need not spend owt [nothing] now need me eye brows threadin to thin pop town tomorrow if fancy it while stayin in to nyt want sum more hand bands xx

Becky: Yeh I need nip town for few bits, go tomorrow then. Let me know wen u ready go, I'm goin gym in morn early as can like xx

Firstly, Becky exemplifies a commitment to partying most weekends as well as hedonistic extensions in the form of house parties, festivals and holidays. This is also evident from her friend cohort who chip in with like-minded comments about these experiences. Secondly, at home, nights out are celebrated by the uploading of the drunken pictures which, in turn, help counter the boredom of the working week; these are the periods when Becky has nothing 'exciting' to say and there is no 'excitement' on the horizon. Interestingly, the Monday status updates are either to say 'Ibiza is coming up' or that 'the job is shit' which suggests a back-to-work blues feeling after the weekend partying – just the sort of feeling the radio stations try to counter – while the Friday updates are more partying oriented. Lastly, and in the context of this work, in the period up to the holiday, the statuses start to reveal the 'holiday hype' around going to Ibiza which, without drawing too many conclusions, seem to attract more social attention in the form of 'likes'. So in essence, the discourses are around mundane home life and excessive consumption at the weekend, and Ibiza. But Becky gives some clues to another feature of the 'holiday hype': shopping and going to the gym.

In the shops, down the gym

Keely: Some of the London clubs are brilliant, yeah. Brick Lane and Shoreditch are like being in Ibiza. The same sort of people, everyone's just really friendly. It's just a different atmosphere, people are just more friendlier and more grown-up ... everyone dresses funky, everyone looks great. It's just like something from a movie. I went to a day rave the other week and everyone was dressed in the funkiest clothes. Big gold earrings and everyone was wearing pink jeans with Converse [trainers]. Everyone just looked amazing, some would be wearing high skirts with funky hats.

Keely is describing the exuberant outfits of people who go out in London but also making a connection that it is a similar sort of style which exists in Ibiza. Many, if not all, in my cohort were planning what they would wear in their holiday destination and often this meant a significant amount of time and money in shopping malls. This was certainly the case for the Southside Crew. Paulie conceded to spending £300 on a pair of jeans for his holiday, £62 on a pair of shorts and £300 on a watch. But this is not exclusive it seems to the holiday. Many of the cohort, such as Jay, are already used to showing off expensive fashion items on a night on the town back home:

Jay: When we go up Southside, you can easily wear £1500 of clothes out. His mate [Nathan] has a £2500 watch and [pointing at Paulie] he is on shitty steroids.

However, the advent of the holiday does prompt new shopping sprees. There is a pressure, because of the social concern about how people will be perceived; they need to appear nice and this acts as a rationale to be more self-indulgent. For example, a few young women in my sample had bought bikinis for every day of their week-long holiday while a group of young men, away for the first time in Ibiza, had invested heavily in the latest Armani and Polo T-shirts. After all, my cohort would say one *has* to fit in to be accepted but I think it is more about that they *feel* they have to fit in to be perceived to be accepted (Chapter 7). This dichotomy will be further explored in the following chapters but the point I want to make here is that as people start to think/talk about the holiday, they feel they need more commodities and accessories to feel part the social occasion and, by having them, that this will thereby boost its enjoyment. Most prominently, as we saw from the latter half of Becky's Facebook statuses and wall posts, this is around buying things even if it means being broke before the holiday has even started. Here is one such experience I observed in my local town centre in the winter months of 2011/12:

As I wander round the Primark aisles, a loud young girl walks past with two others trailing behind as if she is the queen of fashion.

She scuffs in her furry boots which hug close her thick calves which are covered to the hip by those tight legging things which are overlapped by a long-sleeved shirt, a denim and sheepskin coat which is accessorised by a suitcase-size handbag hanging over her arm and large, dangly earrings; the eyebrows thicker than their natural size. She pipes, *'holiday's coming Chels'* as she sees the 'New Season' title above some sandals. She tugs her friend's arm who is dressed in a similar way but has more fake tan and says 'OMG, I've got to get some of these' and handles some £3 sandals. They are thrown in the basket as she says 'got to be looking good for holiday.' [Field notes]

In the course of my interviews and observations, a significant amount of money is spent on these items as it has been constructed for them as the time in which they should 'look good' (Chapter 7). Both sexes frequent beauty salons to get an advance tan, shape up the eyebrows, get a 'Brazilian', wax backs and/or get the nails done. Similarly, many go to the gym in preparation for their holiday; in the main working out at varying levels to get a 'ripped stomach' and big muscles or to shed the flab:

Pamela: Going on these holidays is just a parade of who is the skinniest, who is the prettiest, who has the best pose. It puts so much pressure on you. It is all the comparisons people make. I bought new bikinis and tried to lose weight. No one wants to be the unattractive one, especially on Bora Bora beach [in Ibiza] where you see all nice topless women and ripped guys. There is a pressure to fit in, to feel included. We did it just to make ourselves feel better. We took pictures of pale, white guys, hairy and we had a laugh at their expense. No one wants to be that person. Where were the obese girls? But all my friends are plus size and were so worried about their bodies, dieting and these are older women who should have got over their insecurities. It's a competition to see who looks best, who has the most money. It is really shallow.

Yet there is a whole industry supporting these ambitions at home just so they can look good while they are away for a few weeks of the year. While a few are successful with balancing diets, others cheat and get on the steroids – like Paulie:

Dan: Because you knew you were coming to Ibiza, did you train harder in the gym?

Paulie: Well I couldn't because I've broken my hand.
Jay: I am committed to it. Whatever they say, every meal has been planned because they want to come here cut [looking fit].
Paulie: Thing is, three weeks ago before I come to Ibiza, I think I popped a blood vessel in my head when I was training going 'grrrrrr' [imitates lifting heavy weights].
[Marky bursts out into laughter]
Jay: Your next survey should be male looks and image, and what they are prepared to do to sustain it because in this day and age, they are willing to do any-fucking-thing to get cut.
Dan: OK, what are they willing to do?
Nathan: Men wear fucking make-up now, mate.

Paulie exemplifies the postmodern need for 'men to work at their body and, in doing so, deploy and perform this masculine embodiment as part of a wider construction of identity and the gendered self' (Thurnell-Read, 2011: 979; see also Monaghan, 2001). However, as we know by now, these feelings don't just exist nor are they individually stimulated by the commodity's seductive allure (Baudrillard, 1998) but are also assisted by the persuasive mediums of marketing and advertising. It is no surprise that to afford all this, many of the people in my sample put these purchases (and even whole holidays) on credit – they think about the debt later to enjoy the moment now. So how do these mediums work? A prime example of this is MasterCard's 'Priceless' TV adverts which prompt impulsive spending on credit to feed self-indulgence and a notion of 'do now, think later' – the ideology of credit. In one such MasterCard advert titled 'School's out for summer', which was filmed in Majorca in 2008, numerous teachers from a school frantically rush from their classrooms, celebrating the end of the school year by casting off their official uniforms as they leave. The next scene sees them collectively surge from taxis into Stansted Airport, throwing inflatable toys and suitcases at the check-in staff before numerous planes leave the UK at the same time in different directions. After only 16 seconds, the next shot is from a holiday balcony overlooking palm trees and clear blue skies at which point arrive all the teachers before they jump in a swimming pool at which point 'knowing you've earned every second: Priceless' appears centrally on the screen. There then follows the obligatory information which reassures the audience that MasterCard is accepted in over 26 million locations worldwide. That's a relief, I thought I wasn't going to be able to fully enjoy my holiday without being able to amass more debt.

Home Lives and the Holiday Hype 73

We have already seen how the 'holiday hype' develops its momentum through social mediums such as Facebook and how the ideology of 'being away' prompts people to invest in expensive (or at least more) commodities to appear at their best. A similar sort of process is also witnessed at the airport where the 'holiday hype' reaches its climax.

At the airport

There is further hype as the holiday draws near, which is most favourably marked when meeting at the airport which often signifies 'the beginning of the celebrations'. At the airport, the party begins; couples, groups of friends, friends of friends, stag parties, all converge to mostly easyJet and Ryanair flights destined for places like Ibiza:

> As we walk towards the gate, people overtake; in a frenzy to ensure they get a seat next to each other but would likely have it that they were worried about missing the flight which is delayed. Dressed for night-time adventures, two mothers look on in awe at the hen party they have to manage; there seems to be at least 15 in their crew, all fashioning purple hats, summer clothes which seem to hug their bodies too tightly, all topped with a mixture of fake tan and make-up. They make their way in batches of three toward the flight gate. As I arrive in the queue, in front of me stand two working-class men in their late 20s dressed in designer shoes, twisted-hemmed jeans and matching black T-shirts. As they talk about their drinking action from the Isle of Wight Festival only recently, they compare the new trainers they want to buy on their iPhones. [Field notes]

At the airport, speculative plans are made for the upcoming holiday and stories are also shared, as the 'hype' continues:

> The delay continues and everyone seems to show their frustration. I can feel this is only eating into 'me time', into 'the party hour', into people's leisure time; the time they have worked so hard for and deserve more than anything … When we finally drift like cattle to the slaughterhouse, I see some young men making wanker signs at each other from a distance. They look like an odd group. There is no clear pattern to their outfits and they seem all to be complete opposites of each other. One plays on his Nintendo DS, while another fiddles with his iPhone with his other hand down his jogging bottoms, fiddling with his iballs. After banter between some of them, they half hang off each other as they recount drinking tales. [Field notes]

It is at the airport also that many start drinking in preparation for the 'messy' night (and holiday) ahead. For a few this involves just a beer or cocktail but for others, like the Southside Crew, it involves consumption on another level. In the time before I met them in Chapter 1, they had been drinking 'practice pints of Stella' from 6.30 a.m. in the airport, then around six vodka and Red Bulls and two beers each on the plane. Many in my cohort cannot identify where this ritualised practice originates – though it is certainly aided by the availability of pubs and bars like 'Weatherspoon Express' in the far-flung corners of the airport terminal gates, the tempting offer of alcohol deals on the flights to get people in the mood and even the subliminal messages pasted to the back seats of some of the easyJet flights. As my journey continued:

> I sit only a few rows from the front. Glancing through the literature in the seat in front, I am bombarded with a crude promotion of fragrances, fashion, beauty, sunglasses, travel and many more. Of particular interest are the 'deals' on the drinks: the beers start at £3.60 (€4.50), a glass of wine at £4 (€5), and sparkling wine which is the same size as a glass of wine at £6 (€7.50). The 2-4-1 offers are on wine: two glasses for only £7 (€8.50). The best deal seems to be for half a bottle of cheap champagne which is £16 (€20). The drinks seem to be popular as the steward reloads on Jack Daniel's and has run out of ice after only aisle six which is where we sit. People are obviously in the mood for the holiday and this is confirmed once again when I glance at the easyJet marketing literature plastered to the headrest in front of me which tells me '*to get in the party spirit*': an overt coercion into getting wankered on the flight in preparation for my holiday. It certainly seems to be working as the exotically dressed, hair-braided woman to my right goes through €20 of drinks for herself; a beer, two wines and a Red Bull pick-me-up for the night ahead. [Field notes]

Combine this attitude to getting wasted with celebrating the commencement of the holiday with these messages and the party can get a little too much for some – still the easyJet staff are happy to sell the alcohol even if the clientele are clearly wasted. Later on that same journey to Ibiza:

> Behind me are, from the outset, three female friends in their early 30s; all sinking wine and giggling (two are friends and the other is attending a wedding with her husband). The woman to attend the wedding is 34 and the other two are 33 and 30 respectively; one of

the latter confesses to coming out last June for a *'last blowout'* but returned in August the same year. All seems well until the English one, dressed in a shawl over a bikini, starts to speak suspiciously loudly. The loudness becomes even more apparent when she complains about how expensive vodka is at €4. She summons the stewardess over to slur her complaint. The blonde flight attendant bites her lip as she suggests she can take the drink away, and that the drunken woman doesn't have to buy it. Yet she offers an open-palm of coins to the flight attendant who obligingly picks out the correct money. As they all drink wine, the two Irish women invite the third very drunken one out with them to Amnesia (which is *'not full of skanks* [dirty people or social low life]' says one). The English one is moved by this and launches into a criticism of her husband because he *'is not interested in partying'* and as the women starts to shut down, largely due to the level of alcohol consumption, she starts to lash out at her husband and shouts at him, while flapping her arms across the aisle at him (they are not next to each other but have the aisle between them). The other two women gradually retreat their attention, turning to the in-flight magazine. Shortly after, we land in Ibiza. [Fieldnotes]

Conclusion

This chapter shows that for this group of British working-class young people, home life is governed by what they see as a restrictive social system in which they are expected to study/work or look for it, look after families and have responsibilities. Daily life is perceived to be generally boring and work tedious partly because the social system dents their social mobility, and devalues much of the work available to them and the social credibility associated with it. Collectively, or through some combination, they generally feel this prevents them from 'enjoying themselves' to the maximum and straightjackets their 'freedom'. They say that if they were truly 'free', they would get drunk and/or take drugs 'all the time'. Well, they may be right about their reflections of this 'reality' – how they think about and interpret it – but in general they don't seem to realise that it is ideologically directed through powerful social and media institutions, and marketed through strategic campaigns to get them to buy into a life of spending, excess and 'new experiences' (Chapter 3). These very mediums produce a hyperreal construction of celebrity and the 'good life' which warps the everyday of their home existence so work, looking for it, and looking after family are increasingly seen as mundane. A reality of dreams and fantasy on Facebook or on holiday easily trumps what is perceived to be boring home life (Presdee, 2000; Lloyd, 2013) where the banal pressures of the everyday are seen to be burdensome. This means it has little value other than passing time in it and/or looking for ways to dodge it. One way out is through shopping sprees, beauty salon appointments, watching the *X Factor* and *TOWIE* and excessive consumption at the weekend; it makes their lives, as they experience it, more meaningful, more 'real' and interesting. In the words of a few, they 'live for the weekend' – just as the dominant popular culture would have them believe. And it is this 'live-the-dream' ideology that is reinforced by chat show and docu-reality programmes, magazines, internet and radio shows offering the symbolisms of the 'good life' in the sun, and places like Ibiza become as much holiday destinations as they do places to work. The next chapter considers how Ibiza is constructed around the contours of how this group of working-class British tourists come to learn of Ibiza and what they should be doing there when they go.

6
Constructing Ibiza: The Holiday Career and Status Stratification

Nathan: People say that in Magaluf there is lots of slags and it is easy to pull.
Dan: Is it?
Nathan: Yeah, but people say the birds are quality in Ibiza.
Jay: Nah.
Dan: No?
Jay: No, because they are as good. Right, my mate, he come here [to Ibiza] and had one bird but went to Magaluf and had six birds over seven nights.
[Pauses as if he wants some me to show some commendation of this feat]
Dan: [Almost missing my cue and in a disbelievingly tone] Six birds in seven nights?
Jay: That is definitely better mate, without a doubt.
Paulie: But if you put the work in, you get it out.
Jay: Yeah, but he will put in the work here mate, I guarantee it.
Paulie: But they are like upper class here.
Jay: They are not so easy.
Paulie: Lot more respectable, mate.
Jay: It does get messy here though, mate. The booze cruises are mental. MESSY. 'Come on boys, drinks, drinks, boom, boom' [as he imitates knocking back shots].
Nathan: In Magaluf, it is like cheap man's fucking holiday.
Jay: Yeah but here it is mental, just as messy.
Marky: I don't know actually, I haven't been to any of these places.
Nathan: It's cheap parties in Magaluf but Ibiza is more upper class.

Introduction

The Southside Crew raise a number of interesting issues here. Firstly, they discuss here the way in which different holiday destinations are populated by, as they see it, a certain 'class' of individual; Magaluf in Majorca is where the 'easy birds' are but Ibiza is different because there are supposedly 'quality birds' who are 'upper class'. Indeed, I feel this highlights something quite unique about how Ibiza is constructed which separates it from other holiday destinations around the Mediterranean which the British frequent in that it offers different classes across the social strata the same thing – excessive consumption (Chapter 8). In addition, what the Southside Crew perhaps don't recognise, yet seem to be describing, are groups of young British women from the same class bracket who are more seasoned on the holiday circuit; some of whom have had their fair share of blunt encounters with men and are, as a consequence, less forthcoming to male advances. This, as we shall see, does not deter their candid approaches for sexual conquest (Chapter 9) but the fact they think the 'birds are easier' says something about how young British men in this cohort objectify their female counterparts (Chapter 8). Do bear these issues in mind throughout this chapter but what I want to draw attention to here is the fact that the Southside Crew seem to be accumulating some sort of holiday experience in these resorts.

In this chapter, I want to firstly examine how the holiday developed as a means of time out, and over the last 30 years or so, as a social occasion for excessive consumption. I then want to examine the holiday trajectories which seem to be taking place within my cohort which, to some degree, lay the foundations for further holidays involving the same/similar behaviours as well as annual returns to Ibiza. The main point I raise here is that most of these young people are christened into excessive consumption through a package holiday (predominantly Club 18-30) which encourages them into deviant and risk behaviours. As they mature, and realise that the young and inexperienced are still in those places where they once went, they look for/hear about other places which are similar in their orientation where they can do exactly the same thing but without the young and inexperienced crowd. Ibiza is one of those places and has attached to it a social status for which many aspire. This group have acquired this knowledge through a *holiday career*. However, what we are also witnessing is a younger group who come to Ibiza with no direct holiday experience from other Mediterranean destinations and this is a result of the way in which the island has been made commercially global (Chapter 5). However, for both groups, this is normally the beginning of a series of visits as they realise there are additional forms of status they can claim in Ibiza which prompts them to return. The cultural (Chapter 7) and commercial (Chapter 8) characteristics of this process are explored later but for now I merely want to introduce you to this hierarchy. Pursuits toward this status, I want to suggest, in part help to create a revolving population of British tourists who not only talk about Ibiza, thereby contributing to the ideology of the island (Chapter 5), but more often than not return to spend (Chapter 10).

A very brief history of the holiday

I have already discussed how these British tourists experience home life and how they construct the rationale for the holiday; this is, as I will show in this chapter, something which is both historically and culturally embedded. Where then does the motivation come to engage in unbridled excess and exaggerated behaviour when they go abroad? Here I want to suggest that, with the commercialisation and commodification of the holiday – as a social occasion for time out and the resort as a place to offer this means of escape – has helped cultivate this attitude.

Numerous societal shifts in production and labour have contributed to what we now call the 'holiday'. In industrial society, people worked for 10–12 hours a day then went home. There were 'holidays' but these shrank during industrialisation to be gradually replaced by short blocks of vacation (Standing, 2011). Before mass tourism and the rise in expendable wealth of the lower classes, in the early nineteenth century Britain's working classes holidayed in UK seaside resorts during the summer months. At the time, the introduction of statutory holiday pay meant that the urban working classes could access these sites and engage in conspicuous consumption (Veblen, 1994). However, even then the holiday reflected class divisions as the wealthy stayed in places like Brighton while the working classes in locations like Blackpool (Slater, 1997); some colonisers or wealthy travellers, or members of the upper classes, even holidayed abroad (Meethan, 2001). Importantly, holiday space had particular class dimensions. The seaside resorts thereafter grew rapidly as spaces for leisure consumption which was in line with the development of capitalism (Chapter 3). Yet as we advanced through industrialisation, holiday time became structured around the working life as a means for reward in recompense for hard work. Since the 1950s, the increase in real disposable incomes of the British across the class scale took place at the same time as significant technological developments in transport. This made travelling long distances cheaper and more comfortable (Chapter 4) and this, in part, helped to lay the foundations for the rapid growth in international tourism. However, over the last ten years, the demand for international tourism increased rapidly and has been further complemented with the expansion of cheap airlines which have opened up a host of new destinations (Chapter 1).

There have been further fragmentations to the concept of the holiday with the increasing prevalence of short breaks and independent travel. This increase means tourists have less time and seek to maximise their experiences while minimising the efforts to find them (Meethan, 2001) which means what will be done will likely be of greater importance: the occasion will need to be seized. However, these days, for most, travel and tourism offer an escape from routine and conventional home life. In this world of tourism, everything is commodified: history, culture, heritage and, in the words of David Harvey (2005: 166), a price is put on 'things which were never actually produced as commodities'. Indeed, the culture, tourism and leisure industries are ready to provide an industry around the demands for immediacy and excitement (Ferrell et al., 2008) and we saw some brief glimpses of how this functions on the West End in Chapter 4. These days, holidays have become memorialised as a moment of carefree existence before marriage and parenthood curtails the pursuit of hedonism in leisure (Rojek, 2005).

From this growth in tourism, we have seen two types of visitor emerge: the 'traveller' and the 'tourist'. The former is in conflict with the latter; while the 'traveller' is more associated with middle class notions of tourism and seeks culture and some sort of immersion in local life, the 'tourist' playfully celebrates, and instead seeks a comfortable time away where things are done/organised for them; they seek the old and familiar (Skey, 2011). It is generally agreed that those who take 'package holidays to sites of mass tourism are implicitly poorer, less discerning and lower in both intellect and class' (O'Reilly, 2000: 19) and that a better class of tourist – or 'traveller' – seeks a cultural experience (Urry, 1990). The people in San Antonio that I have studied exhibit O'Reilly's stern description in that they show little adventure outside the resort (unless it is at the Superclubs) and certainly very little, if any, cultural immersion or interest in liaising with the locals.

Despite what may seem to be this obvious class and tourism distinction (Chapter 3), many Britons now try to disassociate themselves from the image of the superficial, fun-loving lower class tourist (O'Reilly,

2000) or 'chav' – someone who engages in *'vulgar consumption'* and lacks *'distinction'* (Hayward and Yar, 2006: 14) – even the ones who are doing practically the same thing as them! This, I want to argue, is partly through maturity as it is an intra-class distinction (Bourdieu, 1984). This distinction I want to firstly suggest takes place through the development of a *holiday career* of excessive consumption among this working-class group, which is normally christened by a Club 18-30 holiday (Hesse et al., 2008), marketed around the concept of youth and the promise of sex (Thomas, 2005) and subsequently reaffirmed year after year in different destinations which offer the same thing – excessive consumption.

A trajectory of excessive consumption: The holiday career

I want to argue that the holiday career is a socialisation process strongly shaped by commercial pressures over the life course as well as situationally in the resort, from which the people in my book have learned about what practices are expected of them when they go abroad and the destinations they should visit to undertake those behaviours. With age, maturity, and to some degree, some experience of getting shitfaced for two weeks on holiday, destinations where many in my cohort went in their younger years like Magaluf, Faliraki, Zante and Ayia Napa are generally constructed in hindsight as crap; they are places where immature people go. Ibiza, however, they have heard, is about style and money so should be their next destination or at least one to work towards in the future. After all, the people who upload their Facebook statuses about Ibiza reap so much social credibility (Chapter 5) and everyone is talking about it. Here it is again. These two young girls from Manchester, aged 19, were away together for the first time in San Antonio:

Dan: Why come to Ibiza? What is all that about?
Gemma: For the 'experience'. Something a bit different.
Dan: But what is so 'different' about it?
Gemma: Atmosphere, innit? Back home I have heard of other places like to go on holiday but not heard ... I mean, people talk more about it, so for us back home, Ibiza is the best place to go, the place to be.

When it comes down to it, Gemma seems to suggest that she came to Ibiza because people talk about it rather than being inherently certain about what or why it means to visit. How quickly Ibiza is reached depends on whether they take *the long road* (learning from experience from package holidays from destinations similar in orientation) or *the short cut* (hearing about it in some form and deciding to go). The former group, I want to suggest, develop this experience through a *holiday career* from other destinations in their younger years which is as much socially constructed as it is commercially directed, and in doing so, mostly likely leads them to the holiday pinnacle: Ibiza. Like Tom and Lee, by the time this group reach Ibiza, they are often in their early to mid-20s:

Tom: It's like the pinnacle, it's like you do Ibiza and you've done it.
Lee: After [I had been to] Napa I couldn't go to Zante and I know if I go to Ibiza I won't enjoy Napa as much cos I'll know that there are better places.
Dan: So are you prolonging the climax as it were?
Lee: Yeah definitely ... By the time I go to Ibiza I reckon I'll have a proper job. It will be a break from doing fuck all. Like for the whole year if I had enough money then I could do everything I wanted in that week and I won't have to worry if I do this, I can't do this, I can just do this, so this. I can completely smash it up.

The latter group, likely as a result of the increasing commercialisation and marketing of Ibiza (Chapter 4), want that immediate credibility such as Gemma in the earlier interview, who seem to jump straight in at the deep end and are generally oblivious to why they have come and how exactly it all works. This group are mostly likely to be in their late teens. Either way, both groups in my sample tend to return to Ibiza, as they realise Ibiza has more to offer them if they have more money – so they have to come back. The few who try to move beyond this holiday 'pinnacle', which is how Ibiza is considered, struggle and I will offer some explanation in this chapter as well as in the next ('In search of the "cultural experience" and the quest for "holiday capital"' in Chapter 7). For now, let's look briefly at these two routes to Ibiza.

The long road

Most taking this route start with package holidays through holiday companies such as Thomas Cook's Club 18-30 which introduces them to what is expected of them to secure a good time on holiday – getting drunk, getting high on drugs, sleeping with as many people as possible, etc. Here Georgina describes her second holiday. She had already been to Faliraki the year before on a Club 18-30 holiday. This time she took a package holiday to Zante. Note how Georgina and her friends exhaust themselves by taking drugs and drinking at the airport and on the plane (Chapter 7) and how holiday reps pretty much encourage her into a cycle of excessive consumption (Chapter 8):

> Georgina: In Zante ... Before we got our plane, we went to Yates [a bar brand name] in the airport. I had had four spliffs before we left and there we had four shots. We had a few drinks, doubles. We got on the plane and had champagne. When we got there, it was the early hours of the morning. We put our suitcases in the rooms and had a drink at the bar, had a look, had a couple of beers and we were completely pissed and laid by the pool and now it is day-time and we have not even started our holiday yet. Then the [tour] reps came round and they were asking people to go on a bar crawl so we went on one that night, we had an hour's sleep after being in the sun all day, got ready and went out.

This initiation led by Club 18-30 often lays the foundations for many to either go back to the same place to do the same thing or look for a different destination, similar in orientation – in that there is a beach, it is hot and there is a zone dedicated to getting wasted – so that they can do the same thing. If they haven't already heard of Ibiza (Chapter 5), it is throughout this process that most start to hear about the island and that visiting it should only be done with 'money'. Take the Beachgirls, who are in their early 20s and said they had been *'around the circuit'* to Malia, Kos and Zante. One says, having been through this that Ibiza is a *'different level'*, while another says *'it's better'*. They agree that those holidays were *'good at that age'* [17–18] and that Ibiza would not have been affordable had it been their first holiday. One jokes that even with £300 they wouldn't enjoy themselves and wouldn't have even *'come out of the room.'* Similarly, take Yazz and her friends who first came to San Antonio in 2008 on a Club 18-30 package holiday after taking their first package holiday in Malia the year before. This time, they also went with Club 18-30. Look how the holiday accessories, which involve heavy drinking and the like, were foisted on them when they arrived at the hotel:

Tracey: Bit of a dive, hotel wasn't nice. It was alright but they was jumping on us.
Helen: Ah, the reps!
Tracey: The reps were buzzing [on drugs] all the time, they are just awake buzzing 24/7. This was 18-30.

Dan: Did they try to persuade you to buy the booze trips and things?
Yazz: Yeah, we did, well we did everything. We picked most things. We went during the closing parties but some things we couldn't do because it was raining, like booze cruise and Zoo Project.
Tracey: But they was really pushy. Come into your room saying 'you have to do this' but we are quite wise but they are pushy. I can imagine a group of girls away for the first time on holiday would just buy everything.

The irony is that they 'did everything' recommended to them by the Club 18-30 reps even though they say that younger girls may 'just buy everything'. In another focus group with some men in their late 20s and early 30s who either were married or with long-term partners and had families, some of them say they have been to Ibiza for some years and the emphasis is very much on making the most of the time they have before going back to what they perceive to be stressful home life situations; they confess to taking on *'fat birds'* if the opportunity presents itself. Not only does it mean they must *'do everything'* to get the *'different reality'* but that other holiday destinations such as Malia – where they went when they were younger – are full of young drunks ... even though they are the ones nursing three-day hangovers having not slept and continually partied as I interviewed them by a hotel pool in San Antonio:

Stuart: You're coming in, aren't you, you know you've got four or five days and you know you've gotta do everything, so ... it's a different reality. Five days to get it.
Dan: So roughly ... take yesterday, for example ... do you reckon you could quantify how much you exactly drank?
Sean: Yeah.
Stuart: No.
Sean: Absolutely annihilated ... no, no, I'd reckon about eighteen pints or something ... easily.
Dan: Eighteen pints? Including these ...? [Pointing as they sit around by the pool drinking pints in the sun]
Sean: No, no ... not been to bed yet ... it's not like that. We're not like the others. It's like Malia's full of chavs.
Dan: Right ...
Sean: We've been to various places like that ... and the general feeling is that people go and get smashed as drunk as they can and get off their tits.

These men are doing the same thing as they did when they were younger, yet are doing nothing different to the 'chavs' in Malia. It is their perception of Ibiza as a high-class holiday destination which separates them rather than anything else. They have come to know it as a place which will tolerate what they do now they are older; now they have been around the circuit. However, increasingly a younger crowd seem to find themselves in Ibiza having heard that it is 'the place to be', having had no previous experience of these holidays. Most of this lot head for San Antonio because it is the cheapest destination and there then begins their virgin experience of excessive consumption abroad.

The short cut

Increasingly, some young working-class Brits short track the *holiday career* and go to Ibiza before experiencing any other destination, which is in part to do with the way Ibiza has bowed to commercial pressures and become mainstream (Chapter 8). Equally, they are also visiting to attain immediate kudos for it. Take this young group of eight 17–18 year olds from Newcastle. Initially, they were unsure whether to choose Ibiza but they knew that people went for the *'opening and closing parties'* and *'perhaps a holiday between'*. Their holiday was to be the sign-off before they ventured into university but what they had come to know about Ibiza was quite detailed: 'It's the party capital of Europe' said one while another said it is *'known to be the best'*. By short-cutting the *holiday career* and missing out on experiences of excessive consumption and learning by mistakes, these young people are at the total mercy of the commercial pressures (Chapter 8); they don't know the rules and learn they are out of their depth. For example, the 'inexperienced lads' from Chapter 5 who were 19 in 2011 were on their first *'lads' holiday'* to Ibiza. Staying for ten days, they only brought with them £700 to spend and quickly got sucked dry of their money, spending €300 to get into a VIP section in Eden, €70 to get into the clubs like Pacha, and even up to €18 for a burger and coke. In short, their reserves quickly dissipated and they had to use credit cards. Even on the beach as I interviewed them, they had 50 euro cents between them to last them until the next day. This is because they had no clue about where to go, what to expect, how to budget, etc. and the consequences of this can place people like them at further risk, as we will see in Chapter 9.

Generally going home feeling a little saddened by these things does not deter people like the 'inexperienced lads' but instead spurs them to return with 'more money' to do it 'properly'. This was also the case for the 'unpredictables' from Chapter 5. They drank two pints at the airport and three vodkas on the plane and went out straight away at 4 a.m. to 'get smashed' (Chapter 7). During their five-day holiday which they organised themselves, they went clubbing every night in San Antonio and were *'wasted'* each night before they were in the clubs, stopping at each bar on the Bay. However, this 'taster' made them aware that there were other places outside San Antonio, other clubs, which they would like to visit next time:

Jerreld: Next time, we will be older, more mature. We will know the right places to go. We will have more money. The girls there love money, so you need that.
Aaron: Some of us were really immature out there so we can do it better next time.

It was, as one said in the last chapter, *'better if you spend more money'*. Also note how they think that the opposite sex will be more attracted to them if they can display more spending power. But these young people see and experience, as well as those who have developed a holiday career, what I call *status stratification*: a way of making this group feel that there is always something better to have/somewhere better to be on the island which will, in turn, give them more social kudos thereby creating social envy and a kind of one-upmanship over the 'other'. The spatial and commodified way in which this is endorsed is explored further in Chapter 8 but I want to make clear here that it is this rainbow-chasing process which keeps many returning year on year, feeling they have to keep up with 'what's in' and 'the new and best places to go'.

Status stratification

When many in my sample encounter Ibiza, they find out there are more ideological social gradings to pursue. The more money one has, the greater social status they can have and this is immensely appealing to this group of people who have come to learn that the be all and end all of their leisure time is to spend/consume to be able to say they have done something or been somewhere where the 'other' hasn't (Chapter 3). They engage in conspicuous consumption (Veblen, 1994) in an effort to distinguish themselves from the people like them in the same class bracket by undertaking things which they consider to be more individual and elitist (Bourdieu, 1984). To climb this invisible ladder which I call *status stratification*, these young Brits seem to make several main negotiations which may also be a combination of (1) finding favour in more hallowed ground in the international resort of Platja d'en Bossa over San Antonio where the 'chavs' allegedly stay; (2) making an upgrade from the basic and cheap clubs like Es Paradis and Eden to more expensive places like Pacha; and/or (3) abandon home life or even tourist status and work in Ibiza for a season (and even then they may have further invisible ladders to climb as 'casual workers'). This is how I think it works.

Spatial: From San Antonio to Platja d'en Bossa (or even Ibiza Town)

As we have seen in Chapter 4, the increasing prevalence of mass tourism over the last 30 years has had a significant impact on the way tourism is experienced in San Antonio, and, perhaps as a consequence, an alternative zone for more international tourists evolved on the other side of the island at Platja d'en Bossa. It is this destination where one can find other British tourists who tend to be either middle class or, increasingly, a group of coming-of-age (after a few Ibiza trips) or seasoned (after years of holidays in Ibiza) working-class 'Ibiza goers'; most of whom have some *holiday career* and see themselves above the populations in San Antonio. In particular, this group distance themselves from the West End and the 'crude' behaviours of the British who party there. They are critical of the behaviours of the younger, inexperienced crowd and see themselves as superior because the people who surround them are 'fitter, more ripped' and 'more classy'. However, as I will show, the only difference is that the price is higher and that they have been made to believe this gives them the right to feel they have one over on the 'others' in San Antonio (Chapter 8). Keely is one example of this group. She has holidayed throughout the Mediterranean and even worked in a resort in Majorca. Since 2008, she has been to Ibiza four times, twice in 2012.

Keely: Well, the first three years I done in San Antonio and it wasn't until I went Platja d'en Bossa this year that I realised how much nicer it is. It is such a nice part, such a better part of the island. Whereas I thought it would be more for the clubbing scene more than anything. It was actually really quite hippified. It was lovely, really laid back and relaxed. It was lovely, it really was, whereas San Antonio has become over the years changed. You're getting a lot more younger crowds and it's getting a lot more like Magaluf is. It's getting really chavvy.

Dan: Why do you think that is happening?

Keely: I think it's because everyone's cottoning on to it. Everyone's hearing about Ibiza and San Antonio and all the younger generations are jumping on it now. Maybe holidays are cheaper now. I haven't got a clue but it does seem to be and a lot of people have said it, it does seem to be a lot younger and chavvier now. That's why, to be honest with you, that's why in July I didn't bother going to San Antonio. It doesn't interest me. The only reason I went in September was because I went San Antonio Bay for the hen do and I hated it. When I went back to San Antonio from the three nights I was staying in Platja d'en Bossa, my friend had a table booked for Mambo. So I literally went to Café Mambo just to watch the DJs then I went home. I didn't even actually go into the West End. I wasn't interested.

There is a clear divide in the narrative yet she has much in common with the people she is describing in San Antonio – even though she views the resort as a location where the 'chavs' go which is now no different to other tourist zones across the Mediterranean like Magaluf. Still, she goes to the area to appear to have one over on the 'chavs' by booking a table in the new, plush bars of Café Mambo while the rest just drink on the rocky beach below (see Chapters 8 and 10). Keely thinks that 'the young people have invaded' San Antonio, and maybe they have because of Ibiza's commercialisation, but perhaps also it is that she has just 'got older', and in doing so, feels she has grown tired of the relentless activities associated with the West End ... or maybe not. I only say this because Keely still drinks copious amounts of alcohol and takes MDMA in Platja d'en Bossa and the nearby club Space. It's the same for this woman in her mid-30s whom I met on another flight to Ibiza:

Avoiding the overpriced easyJet snacks, I sneak out my food and drink from my bag and start to eat on the flight. I strike up a conversation with the woman next to me and it does not take long for her to completely launch into a dialogue about herself. She is staying in a small village outside Ibiza Town. When I ask her about the 'no-go' areas in Ibiza, she immediately says *'San Antonio'*. As she says it, she sneers and encourages me to divert myself from the area. Later in the conversation, however, it transpires that she used to work as a PR girl on the West End. She has done *'all the clubs'* and has *'taken all the drugs'*. *'Not my scene any more'* she confesses. However, she also concedes to avoiding *'Bora Bora, because of the sweaty, drugged-up people'*. She smiles again as the wrinkles come alive on her face while large globules of mascara hang from her eyelashes. [Field notes]

People who return to Ibiza seem to have taken a rite of passage which, for most, starts in the resort of San Antonio, extends to Platja d'en Bossa, and then perhaps to Ibiza Town or a quiet village/private villa. Another way to obtain status is to frequent the Superclubs while, at the same time, show some spending power.

Social: From Eden to Pacha, and from beach to 'beach club' (or even beach club hotel)

The Superclubs are considered one of the main reasons why young British, as well as international tourists, come to Ibiza. However, the price to enter these clubs varies significantly: there are days when ticket sellers almost beg people to enter Es Paradis for free on non-themed DJ nights while attendance at somewhere like Pacha, where a higher class of individual is considered to go, requires at least €70 for the night (depending on the DJ) and of course drinks are on top of that and taxis home are also a consideration. When many come to Ibiza, they encounter this ideological invisible ladder and often return because they want to experience the other clubs just so they can tell people they have '*done it*'. For example, this young group of lads who have been to Malia and Ayia Napa say that 'everyone says that Ibiza has the best clubs', like the idea of its wealthy image and want a piece of it:

Harry: Ibiza is the place where the wealthy go, the Superclubs, Amnesia, Pacha, Privilege' – bit of both. They would never bother going Malia or somewhere like that. That's for really young people.
Dan: So it's a bit of both worlds. Different ends of the social strata. And you're somewhere in the middle?
Harry: Scraping through to the middle.
James: We're at the bottom but we're trying to get up [laughs].
Dan: You want to look good and afford it.
Harry: Tried going to the gym but I gave up, but people here do seem to care more about their appearance than in Malia. You see more ripped men and nice women on the night out.

There is no mention of Es Paradis or Eden but the bigger clubs. They even concede that by engaging in conspicuous consumption (Veblen, 1994), they hope to make a class upgrade but then seem to remember who they are at the same time. However, after reaching San Antonio and spending some days there, they realise that they are on the bottom rung of an ideological ladder; they need, it seems, to start again and the only way they can do that is if they have money, which means they will have to come back – because Ibiza is something which has do be done *'properly'* in life:

James: But obviously you get people here who have got the money then you get people like us who haven't got the money but want to be here and they try and do it on a budget.
Harry: They wouldn't be round here anyway, they would be in villas and just go direct to the Superclubs.
James: To be honest, I didn't think I would be 22 and in Ibiza. I thought I would be like 25–26.
Dan: Why?
James: I wanted to be able to do it properly.
Dan: How do you do it 'properly?'
Harry: In an ideal world, I would want to go to each of the clubs at least once and drink and not have to worry about how much it cost. I hate scrimping and scraping to get a couple of drinks.

Indeed, James had to borrow money from friends to afford the holiday and this is frustrating – especially when he can see how participation in the 'Ibiza experience' requires a significant amount of money. Only the night before, after paying €45 for entry into a club, he spent €85 on drinks alone. Even in this short exchange, it dawns on them that even in their early 20s, they don't quite have the money to 'do it properly'. They want to reach a stage where they can look frivolous with money and spend it as if it has no value – just like the elite (Veblen, 1994), just like the celebrities (Chapter 5). The next year they vow to return. It's the same story for many young British working class; they hear that there is a better club, somewhere better to go and this is often enough for them to save up and do it 'properly'. Alternatively, they could try it and feel/look stupid by not having the expected spending power, like this young man I met on the West End one evening:

We walk up to the West End and take a seat to buy a drink; the offer is €10 for two drinks, a shot and a jug of cocktail between us. It is about 2 a.m. We sit there trying to absorb the theatrics in front of us. My chair is conveniently close to another young lad next to me and we share the same elbow space; it makes it easy for me to strike up a conversation with him. He is a young working-class Londoner, slightly taller than me and definitely more fashion conscious, as he wears a bling Gucci watch and American baseball cap. He says he went to Ushuaia [a beach club hotel] the other night and thought it was '*amazing*' but too expensive. He paid €70 to go in and see Swedish House Mafia and had two drinks but couldn't afford to stay. When he described the experience, he favoured the architecture inside. As we continue to talk, he reflects at how Pacha was '*snobby*' but thought Eden was alright and that he would probably '*end up*' in the latter as the evening progressed. However, he doesn't rule out Pacha, saying he would need more money to do it '*properly*'. [Field notes]

For the time being, this young man is comfortable with 'ending up in Eden' – one of the cheaper Superclubs – although he looks as if he could belong in another because he has all the kit (Chapter 5). He has had a taste of the 'good life' in Ushuaia and liked it, and even if Pacha was 'snobby', he concludes that the only way to appreciate the Promised Land is to have more money to spend (Chapter 8). It may not all be about money, for status is also gained by becoming a casual worker.

Structuro-situational: From tourist to casual worker

'Casual workers' – such as those selling tickets, promoting (PRs), dancing, working in bars, or the like – see themselves as a cut above the tourist (O'Reilly, 2000) because, as they see it, they are the ones who are really 'living the dream.' Most have abandoned, what they perceive, to be a shit and uncertain home life (Chapters 3 and 5) and are instead surfing the wave of their youth and freedom in San Antonio. However, they are not just living and working in Ibiza, but most are pretty much engaging in a non-stop 24/7 party which although is detrimental to their health and wellbeing (Hughes et al., 2009, Chapter 9), is something which carries a significant amount of social envy – even if it is 'dirty work' and the pay is shit (Guerrier and Adib, 2003). Casual workers have the inside know-how on where best to go for the sunset, for the best quality drinks at the best price, to get the best drugs, and often get invited to private 'workers' parties' and even get on guest lists to clubs. The tourists look up on the casual workers and envy them because they have access to this party lifestyle on-tap while the casual workers look down on the tourists because they are only here for a short period of time and don't have the balls to do what they are doing. It is not easy to get on the casual worker ladder either; having heard about 'the dream', the novices often turn up unprepared hoping to get work and, in the process, attain some social credibility, and believe me, people are queuing up to try and realise this 'dream':

> I meet a young woman with bright, reddish-purple hair among two others with bleach-blonde equivalents. They all wear *à la mode*

80s-style, retro clothes and Ray-Ban sunglasses. Among the three is one of the three ticket sellers I met on the first day of fieldwork [four days ago]. When we first talked, she was hopeful for a new job selling booze cruise tickets and/or club tickets. They look forlorn. Two just stand there and after nearly a week here have got nothing to show for themselves. One of the three has managed to get an interview for a booze cruise job. The interview process requires her to 'survive' a booze cruise to show she won't go crazy with the drinking and it seems her performance will be assessed like this. She seems hopeful. [Field notes]

The irony is that the 'job interview' is a test of her ability not to party while a party is going on around her. I have already discussed the way in which these British tourists are exposed to the potential lifestyle of working in a resort through the various media and social institutions which act as a blatant advertisement (Chapter 5). Indeed, Ibiza, as well as other Mediterranean destinations with a reputation for excessive consumption, becomes a prime destination where this 'dream' can be played out: after all, everyone sees how people like them on *X Factor* are chasing their 'dream lives' as well. In fact, there are even 'casual worker' websites and blogs which also offer up the glitz and glam of a job in the sun. Some even charge up to £70 to help young Brits – no doubt out of work, between jobs or studying – to look for work abroad. While some young people in this kind of precarious transition bracket (Standing, 2011) may be tempted to ditch their jobs back home or between university semesters come out for a season, they may equally be tempted, after only a few days in this adult playground, to make an impulsive decision and make the holiday permanent by looking to work a season in Ibiza. Here is one typical example:

We manage to find an internet café on a side street from the main drinking strip. While my colleague checks her Facebook, I hang around and see two slim, young British men come in, both looking a little worse for wear; one carries a plastic bag with him. They go up to the bored Spanish woman at the desk and ask how much the internet costs. *'20 mins for a euro'* she says and one of the thin figures starts to fish around in his pocket but among the bronze coins cannot find much. I suggest they use the internet after us as we won't need all the time as they don't seem to have money between them. They appreciate this and offer me a cigarette as we wait to use the computer. We lean against the building, smoking while the evening gets under way on the West End and people walk past in their best summer gear. It soon transpires they have been here for three weeks and are both working out here ... until one, Greg, confesses he is still *'looking for work out here'* as he quit his commission-only job because it was *'shit money'* [earning €25 a night yet accommodation costs €20]. He came out having quit his job in the UK. The other, Rob, has a job selling boat tickets for Pukka Up [a notorious booze cruise which leaves from the Marina] and Stormbabes [another similar cruise where topless women dance around to music]. At the moment, they are sleeping on the beach to save money for an apartment and have their clothes at a friend's apartment. When Rob says he has spent over £2,000, I choke on my cigarette but then he reassures me he has his 'ten grand student loan in the bank' but 'won't be going back to uni now' [because he has decided to work in Ibiza]. They say Es Paradis and Eden are boring but then boast about their escapades in *'Pacha the other night'*, saying how I would like it because it is full of *'older people'* at which I take no offence and laugh. [Field notes]

Note the precariousness of home life (Chapters 3 and 5) and how the attraction of living/working in Ibiza seems to have helped sideline the ambition for university (Chapter 10). But the brutal truth is that it is not even that easy to 'live the dream' because there are so many like-minded others seeking a similar utopia. The work they do barely pays for them to get by on a day-to-day basis, is just as uncertain if not more than what they were doing in the UK – but this is part of 'living the dream' isn't it? In this next excerpt, after explaining how they use MCAT and Ketamine most nights, these young PR girls reveal the difficulty they have had in working their first season in Ibiza. Again see how the precarious home circumstances were, in part, one reason to come out to Ibiza. Yet the move has not necessarily benefited them:

Dan: What do you do?
Paula: I am a PR girl. That's what I mean, so many people are off their head [on drugs]. I am telling you from experience.
Dan: So why are you out here?
Paula: I wanted a change. I had too much shit going on in my life back home so I thought I would come out here and have a laugh and sort my head out.
Dan: How are you feeling now about the job?
Karen: I do it as well, you know.

Paula: I work 8 or 9 hours a day, walking up and down the West End. Some days you make €10–15 a day.
Dan: So they don't pay a straight wage?
Karen: No it's just commission. Some do but the place we work, we don't get it.
Paula: They give you drinks but that's not really a wage because if you are pissed, you just can't be arsed talking to people. That's why I don't drink when I work.
Dan: Is your rent expensive here?
Karen: Like €530 per room. There are five people here. It was meant to be €290 each but we got ripped off. It's not good where we are staying.
Paula: We got ripped off.
Dan: Is it worth it: to live and work out here?
Paula: It is in a way but I wish I had known more about it. I had to save for a deposit and put that down. Didn't have a clue about that. I didn't have a clue about National Insurance numbers. We had to get one which took five weeks.
Dan: So you have had to learn the hard way.
Paula: Yeah and next year we will know better. If you feel sick, you go to the doctor's for €70.
Dan: So what's the plan? What next?
Karen: Don't know, work in McDonald's.
Dan: So it is worth it?
Karen: I don't want to go home. I love it here. It is my home. It is proper work.
Paula: It is not reality though.

The narrative shows that the PR scene is rife with drug use and research shows use is substantially higher among PRs than the tourists (see Hughes et al., 2009; Briggs et al., 2011b). Also witness how wages are subsidised by drinks because they are 'living the dream' in exchange. However, this mix makes for an even more precarious predicament as they place themselves in a lifestyle which involves almost no money and little direction. Still, although they knew little about how to do it, and in the process had to learn the hard way, this does not deter them; far from that as it acts as the catalyst to come back out and 'live the dream' *properly* next year. What happens between is irrelevant – such as menial or even derogatory work – because the end goal is more to generate social status, and in the process, create social envy about what they are doing (Chapter 3). However, in this way, they confine themselves more in the temporal and make their pathways even more uncertain (Bauman, 2007).

These young women are not the only ones. During the summer of 2012, I was told by the local police of San Antonio that they had 3,000 registered casual workers yet only 1,000 had jobs. The rest? Probably working in the informal economy to sustain the 'dream' and perhaps unsurprisingly this is why many end up augmenting their poor wages by selling drugs as well (Chapter 8). These ambitions for a higher status derive in part from an individual need to appear better than the 'other', to be able to say they have done something which 'others' in the same arena have not. The other side of it is how they are commercially duped into thinking this (Chapters 5 and 8). However, despite all these status shifts, many end up doing the same thing (going out, getting battered drunk, taking drugs, spending money in strip clubs, taxis) and not only does their money continue to go into the coffers of the rich club, bar or hotel owners but stays reserved to the resorts, the Superclubs, the plush shops and expensive beach clubs.

Conclusion

This chapter has shown how this group of young working-class Brits are gradually socialised into believing that the holiday encompasses particular behaviours and it is these actions which they reproduce in various European destinations catering for their holiday ambitions for a blowout. There is general agreement that Ibiza has it *all* and is the *best* place to go. It is culturally relevant in home discourse and forms part of the self-rationalisation to enjoy the 'good life' and/or 'live the dream' at the expense of perhaps uncertain work or future prospects (Chapter 5). A significant number of people in my sample, when they are young, learn of Ibiza through media, news, friends, and normally try a Club 18-30 organised package holiday to start with in another destination where they are shown exactly what they should be doing on holiday (drinking, taking drugs, sleeping with each other, taking risks). This, for some, lays the foundation for subsequent holidays in different destinations which offer the same thing – a space for excessive consumption – and often result in the same behaviours. They embark on a *holiday career*. When this group find out there are 'better places' such as Ibiza which with their attendance bring them more social kudos and status, they start to save accordingly to be able to do it 'properly'. However, the chapter shows that some of these young Brits are short-cutting this career and jumping straight at Ibiza without the resources or experience; largely because of how aggressively the island has been marketed in recent years. Once in Ibiza, or having had some experience of it, a form of distancing and distinction (Bourdieu, 1984) seems to occur on a number of different levels: particular places, clubs and behaviours generate a social frowning from those who have come to know Ibiza as this group find out that more spending equates to more status. Importantly, however, the market recognises this need for status, and takes full advantage of it; that is, it's not because they *are* better but believe they are better because they *can afford* better things in places which supposedly look better (See *The spatial commodification of status stratification* in Chapter 8). But there are further subjective individual rationalisations which buttress the deviant and risk behaviours of this group and, in the next chapter, I want to explain what happens to this 'self' with the advent of the holiday and how and why they become rationalised as permissive.

7
'You Can Be Who You Want to Be, Do What You Want to Do': Identity and Unfreedom

Jay: That was how we were in Magaluf mate, we couldn't remember the first day. Maybe it will be like that today; well we can remember everything so far. I thought I would be more minging by now because me and him [Nathan], we don't really drink.
Nathan: That's because I have a girlfriend.
Dan: You have a girlfriend.
Nathan: And a kid on the way, mate.
Dan: I see.
Nathan: But I am sexually frustrated, mate.
Jay: Yeah, sexually frustrated chump!
[All laugh]
Paulie: I've got a girlfriend ...
Nathan: But mate [grabs my arm], what happens on holiday, stays on holiday.

Introduction

The Southside Crew raise some important issues here with regard to how the holiday marks a shift in identity. Firstly, Nathan feels he can finally do what he wants to do – he is liberated somewhat. Secondly, and aside from the fact they can't remember the first day of Magaluf because they were so drunk, the holiday is constructed as time to do the things they feel they are missing out on as well as the things they can't do at home. For Nathan, the holiday – and what he does on it – bears no reflection on his home relationships and the new baby which is due soon; there is also some clear determination in his tone to take advantage of this moment because no one will know who he is and there are no apparent repercussions.

In this chapter, I want to discuss what happens to the self as it decouples from the perceived restrictions on everyday home life, routines and responsibilities into the supposed liberation of the holiday scenery. The first thing this does is create the immediate impetus for most to initiate default behaviours and practices which they would normally do at the weekends (Uriely and Belhassen, 2006; Tutenges, 2012): that is, get drunk, take drugs and engage in playful forms of deviance and risk-taking (Chapter 5). Yet the people in this book are seeking to maximise their experiences in the short time they have to do it (Meethan, 2001) – it may not come again so the occasion must be seized and drained for as much as possible, much like extracting the fading notes from a concertina as it is exhausted of all its air. It is therefore in the excitement of this 'liberation', heightened by a group dynamic which endorses drinking and drug-taking, general experimentation and a subjective intention of self-reward and indulgence (Chapter 3), that the first night is often the most risky and is marked by most by a margin of excessive consumption into the realms of the ridiculous. I would like to call this level of excess *hyperconsumption*.

While this is partly because Ibiza is depicted as a place back home where one can engage in these behaviours with no visible repercussions (Chapters 4 and 5), the other side is that this decoupling means people are increasingly likely to try things they would not normally do in a place which is constructed as unfamiliar (it is a foreign country and a plane has taken them there) as it is familiar to them (in that they have come to know what is expected of them and are in an environment which reflects home NTE symbolisms). In this sphere, a *new permissiveness* is constructed around anonymity which means they see no direct reflection on what they do as 'wrong' because everyone around them, including their 'friends', are doing the same/similar things and there is little pressure for them to adhere to their daily demeanour back home. This is especially true when the social context is offering the 'good life' for which they have come to be so familiar (Chapter 5). In San Antonio, my cohort are made to feel comfortable through the familiar pub names, bars, take-aways which reflects what Billig (1995) calls 'banal nationalism' – the everyday representations of a nation which provide an imagined sense of national solidarity and belonging. I will address the role of the resort in this respect in Chapter 8 but here I want to discuss this identity transition which alerts the individual's subjective sensations to this *new permissiveness*.

In doing so, I want to start to challenge notions that these behaviours represent 'freedom' and 'unconformity' because from what I can see they represent only exaggerated forms of excessive consumption or *hyperconsumption* which typically take place at home, and even extensions of these behaviours into very unpredictable realms (experimenting with deviance and risk), instead symbolise efforts to escape *unfreedom* and *conformity* – this being their commitment to consumer lifestyle and the way in which they are bound by reproducing the same practices typically undertaken on a night out back home (Bourdieu, 1984). In a similar vein, because much of what is going on around them is the 'same sort of thing', in the 'same sort of places', with the 'same sort of people' and everyone around is wearing 'similar sorts of clothing', this, I want to argue, prompts subjective adventures to claim back 'individuality' which can also reap social commendation. These are additional elements to consider in the process of engaging in extreme deviance and risky behaviours; it is leap of faith into the unknown which, for some, becomes legendary but for others, can go very wrong (Chapter 9). I begin with examining how a *new permissiveness* on behaviour is constructed with the advent of the holiday.

Identity

The attitudes which underpin the deviant and risk behaviours of these young Brits represent a self-rationalisation that this social moment (holiday) and new territory (resort) – the time and space respectively – offer a *new permissiveness* which the home turf doesn't; it allows for the loosening of moral boundaries because (a) people are temporarily relieved from home routines and responsibilities; (b) no one of real authority is telling them otherwise what they can or can't do; (c) everyone around them, including their 'friends', seem to be doing the same sort of thing; (d) they have come to learn that this is what should be done and is expected of them (Chapter 5); and (e) the location is familiar yet unfamiliar; 'familiar' in that it has all the symbols of fun and excitement which they recognise with yet it is 'unfamiliar' because it is in a foreign country. So a *new permissiveness* on behaviour is constructed which is structural, spatial and situational. As part of this transition, *time folds*: the young Brits revere in the ideology that this is what they should be doing with their virgin youth while those escaping home life responsibilities, who are slightly older and are more likely to have families, equally fetishise the opportunity to seize the moment to 'get on it'. It is this subjective layering which provides the foundations for engaging in deviant and risk behaviours.

Time folding and a new permissiveness

The broad attitude to the holiday; that in this short window in time and space, everything must be claimed and must be done in a way which meant that enjoyment was had at every single moment (Meethan, 2001). Much like the MasterCard mantra, these moments are 'priceless' and must be seized. As we have discussed thus far, the social construction of Ibiza exists around an ideology of 'something which *must* be done in life' (Chapter 6) and much of this discourse dances around notions of 'youth':

Cathy: It's somewhere where you know you have to go when you are young.
Dan: You have to?
Anna: Well, it's something to say you have done. To say you have been to Ibiza, it is an experience.
Cathy: You see other people have been and other people are going, and everyone is talking about it and it is like you have it in common with them.

In the process of my questioning, look how the ideology of Ibiza of 'somewhere to go' or something which 'has to be done' becomes something to have 'said which has been done'. This is because going there and doing these things retain very little personal ontology and instead, as Anna says, is more reflected in the discourses of Ibiza. There is no actual substance to Ibiza and, as a consequence, it gives nothing of essence to the self, only social credibility and the security they have done it *'when they was young'* because that's what is expected of them. Similarly:

> Approaching the queue at the airport gate, a pair of young women catch my attention. They are both wearing noticeable T-shirts and beckon another two to the queue. Without much thought (and plenty of cheek), they join them. The queue is going back and forth on itself so we pass each other several times. On their T-shirt is symbolically written on each breast 'Magaluf 2010' and 'Ibiza 2010'. They look young and only later tell me they are in their early 20s. As we pass them once again, I lean over: 'So tell me what's the deal with Magaluf and Ibiza?' Turns out they are flying to Ibiza for five days and then Magaluf for seven. The whole trip, they tell us, cost them each about £430 which is cheaper than a package holiday. Their boyfriends are flying out to Ibiza separately with their friends.

I ask why they don't go out together and they reply that they would only *'cramp their style'*. One says 'it would be a different holiday. If we wanted to go away together we would like go to somewhere interesting like Egypt'. They smile and laugh, and toss their bleach-blonde hair from side to side. On the plane, I sit relatively close to them, I approach them and ask them if they would mind having a chat about their holiday to come and the expectations they may have. Three of them end up answering the questions. They each have nicknames for each other like Bam Bam and Toiletmouth and seem to feel quite happy to call each other thus. Back home, they go out to 'wind people up, [have] dance-offs, have a laugh, mess around'. They have with them £800 for the ten-day holiday in Ibiza and Majorca – £500–£600 they estimate will be on Ibiza.

Dan:	That's a lot of money to spend.
Toiletmouth:	Yeah, but we only go on holiday once a year, nah what I mean? Think of all the memories you are gonna get. You have no kids, no commitments, you won't be able to do this soon.
Bam Bam:	Yeah, it's like the memories. Like the next day we go through all four cameras to see what happened.
Toiletmouth:	We wait all year for this as well. We have to make the most of it.

Many in my sample, like Bam Bam and her gang, reason that while they are young they need to take advantage of their youth and do all the things which are expected of them (Chapter 5) so that it can be reflexively visited later in life. When they look back, they can say they did as many 'crazy things' and had as many 'experiences' as possible when they were young; because there will be a time in the future in which they will have to be responsible. This doesn't mean that even when they reach this age or this point in their lives that they will not go back to that time meaning that 'youth' is not the only rationalisation. Some in my sample already have responsibilities such as jobs and families and perhaps are looking to take advantage of things they felt they 'missed out on' or 'deserved in any case'. So the *new permissiveness* in identity is constructed around where people think they are in their lives. This is either:

1. 'I am young, the future is uncertain and this is what young people must do to enjoy their youth so we better do it before we get older and get responsibilities'.

Or:

2. 'I am a bit older [or maybe not] but have all these responsibilities in my life and so deserve time out and need to make the most of the short period of time I have away from a pressured life'.

In fact, a few people in their late 20s and early 30s in my sample said they *'looked forward to the next stag party'* and one said the *'next golfing holiday'* because it signified a similar opportunity to escape the home tedium. These are potentially the opportunities which may replace Ibiza later on in life and this is precisely how these people move through their lives these days: constructing opportunities to be away from lives which they think are boring and/or pressurised. The important thing to point out here then is that Ibiza is sold to them as something essential to the 'youth' experience as it is to revive 'nostalgia' (Chapter 10). The *new permissiveness* – made available through this window in time and space – is thus constructed as it supposedly momentarily annuls other home concerns and responsibilities and this is reinforced, to some extent, by the anonymity of the self, the social relations and the new environment.

The anonymous familiarity of holiday relations

This *new permissiveness* which is constructed around behaviour functions under the umbrella of a shared anonymity in the holiday context: people know they are away from their home club nights, and although they are often with their friends, will encounter new random people in an environment which is not the town centre back home (although it displays all the hallmarks!). The fact that they are away from these familiarities means that what they do is not under scrutiny from others whom they may know and this means that what they do cannot easily be reflected back on them as a symbolism of their home characters (unless they make them public on Facebook or otherwise). It is this which helps to allow the self to sanction some experimentation with behaviour beyond this home image. It also contributes to a loosening of home identities because keeping up appearances is not on the 'going-out agenda' but rather the innate satisfaction of personal fetishes and fantasies, and so anonymity guarantees this transition. Take Helen, for example who says at home she is *'quiet'* but the holiday allows her to become *'someone else'*:

Helen: I don't know, there's something about the sun and before I go on holiday I work out, I buy new outfits and you have a tan.

I just feel really confident when I go on holiday and I'm a lot more relaxed away from the home pressures. I come from a small village and when I'm away I'm not near the village gossips on holiday, no one's interested in what you're doing, everyone's just letting go and having fun.

Helen recognises that the fact that others around her have the same attitude was some measure on this identity loosening. However, when she came back from her first holiday with her girlfriends in San Antonio, there were regrets; probably because the sensible and measured character she says she is at home was taking a siesta while a darker form of her being – which she blames on the alcohol – was allowed to fiesta and explore the persuasive margins of what is expected of her on holiday and what she may subconsciously expect of herself having learned about it (Chapters 3 and 5):

Helen: To be honest I really regret it. I was really drunk and I ended up waking up in bed not knowing what had happened with this random [person] next to me, I mean it's not how you would choose to remember your first time [or not remember it]. I know people say drinking isn't an excuse but when you have no memory of what happened it obviously plays a part. I came back a non-virgin with an STI from sleeping with one guy in a drunken state. I definitely learned my lesson not to do that again, it's dangerous and anything could have happened to me but I obviously couldn't handle all that alcohol I had drunk; most of the night is a complete blank.

Firstly, Helen attributes this to the alcohol because this is the dominant ideology surrounding substance use: that alcohol *does things* to people rather than it is what people *invest in* alcohol. This pacifies her actions and makes her feel responsible – and this is just how the social system would subscribe it (Chapter 3). Secondly, she cannot identify how the social occasion, and more to the point, how the social expectations of her, came to fruition and merged with the commercial nature of the environment to put her in this risky situation (Chapter 9). This is the power of the ideology. In another example, this group of young men, who paid for a *'random'* man with no money to get a blow job from a prostitute and beat up someone on the beach *'for a laugh'*, pretended to be Queens Park Rangers[1] footballers and got into some clubs as VIPs and as a result *'got loads of free drinks'*. Here they reflect on cheating on their girlfriends while in San Antonio:

Charlie: 'When the cat is away, the mouse will play' – I was in a relationship when in Ibiza but 'having fun'. It's not that I don't love her but I am young and stuff. How can you be in Ibiza and not live it to the full potential? There's girls there walking around in hot pants.

Aaron: I saw less than that, boy.

Other boys: Yeah, yeah, yeah.

Dan: What you're saying is to fully experience Ibiza one has to or is expected to ...

Charlie: [Interrupts me] Well in my opinion, yeah. I had a girlfriend, it wasn't that I didn't care about her but I just felt like I am in Ibiza, I am gonna have fun.

Dan: You thought 'I'm hot, I'm young and I have a nice haircut. She'll never know.' Do you feel bad? Be honest.

Charlie: A little bit if I am honest [he slept with five different girls]. It was not because I was drunk. I hate people that say 'I was so drunk'.

Note how 'youth' and the 'expectations of youth' as well as *being* on holiday are part of the rationalisation for the cheating. Secondly, the social construction of Ibiza as a destination where one should do these things is evident in Charlie's narrative and it is this which also becomes the rationale for cheating on his girlfriend. However, they seem to suggest that 'how could one not do it' when confronted with the green pastures (or maybe fag-butted beaches) of semi-naked women, nice weather and alcohol – much like Liam and Graham in Chapter 6. The holiday prompts the identity construction of 'someone else' and it is this 'someone' (or the alcohol) on which many in the sample place responsibility for their misdemeanours on holiday and, to some extent, allows them to think they go home guilt free. It was 'someone else', doing 'something' which was distant to their character and it was done 'somewhere else'. This is why many say things like 'What happens in Ibiza, stays in Ibiza,' but this is a thin rationalisation because their friends undoubtedly learn of their escapades (and there is always the danger that the stories or photos could find their way to Facebook). Here, Sarah starts to suggest this is the case:

Sarah: Like, whatever happens in Ibiza like, it's not gonna go back to England. What happens in Ibiza stays in Ibiza, so you can do what you want but you don't have to pass it home.

However, later in the interview:

Sarah: Well, you probably will [take it home with you], it probably follows you back anyway, but it's like on holiday, you get wild and you just do what you want and you can't do that in England, cos you know everyone and everyone knows everything, but when you're here, like with the girls and mates, they don't budge you so you do what you want, yeah?

Perhaps, to some degree, without the familiar weekend crowd in the clubs back home to sustain 'who the person is' on a night out, new attitudes and behaviours can be initiated which are socially reinforced by the cultural field of the resort. Yet there is a tension here: Sarah would be likely to go out with the same mates at home as she would on holiday, wouldn't she? Therefore it is not necessarily the immediate social relations around the individual on holiday which counter identity transitions, self-exploration and experimentation but the ways the groups recognise/permit that the holiday moment offers a *new permissiveness* and the way the new space is rationalised as a vehicle for this shift. At home, they may not be so forthcoming on exploratory/exaggerated behaviour for it may affect their local reputation in the long run. But the holiday time is temporal carnival so the social rules are different (Bakhtin, 1984; Featherstone, 1991). In addition, living this fantasy life for such a short period encourages the weekend home self out of the woodwork into an everyday routine; the weekend 'person' can *be* on an everyday level and the more this routine develops, the more the self starts to see itself outside itself (reflecting on what goes on at home), which for most, prompts irregular and redundant feelings about any potential malaise in their home lives.

Self-deconstruction

It is in these moments of living the 'good life' in San Antonio that most encounter a daunting self-deconstruction. Out of context, and in and around a plastic splendour of bright lights, bars and clubs, and unlimited hedonism, home life suddenly appears oddly out of kilter; the 'then' (at home) is looked back on as more oppressive as the 'now' (on holiday) becomes the liberation. As we shall see, the 'will be' – which is returning home – becomes reconstructed as even more oppressive as the holiday nears its conclusion because it swallows up the 'now' or the moment in which they can truly *be* themselves (Chapter 10). It is in this period that a self-deconstruction takes place on an individual reflexive basis which is bolstered on a social level between friends as they ponder home life. This leads most, I want to argue, to artificial conclusions about themselves and their home predicaments (Chapter 5) because on holiday, the mind starts to linger on the supposed shitness of home life as conversations get deep as the time with 'friends' starts to open up personal channels. This, in turn, triggers reflections and a kind of life evaluation. Take Pamela, for example here. In this exchange, we discuss how this self-deconstruction takes place:

Dan: So if you are reasonably happy before going away, why go away on holiday to 'escape'?
Pamela: Yeah but at least you don't have the bills and other people are waiting on you. Like sometimes I have been in a bad place and booked a holiday on a whim ...
Dan: Because you thought, 'fuck it'.
Pamela: Yeah, fuck it I am miserable, need a holiday to cheer me up but I come back depressed so the holiday was a bad idea but I have been on holiday. But you can't do it like that because all the worries you forget, you have to pick up again and they come down on you hard.
Dan: Is it heavier?
Pamela: They can seem heavier. You juggle them at home as they are not always in your hand at once so it's never that bad but once you drop them – you go on holiday – you have to pick them all up again. You have to start again and get yourself back into the momentum of dealing with them. At home, you don't have time to think properly about yourself and what you are doing, you can't analyse as much.
Dan: The juggling balls right, so there is no time to look at life.
Pamela: You are seeing everything differently. It is re-evaluation.
Dan: Why would you not think about those things? Why deny yourself of those thoughts?
Pamela: I do anyway.
Dan: Are you not thinking about your worries on holiday at all?
Pamela: Well we do talk about worries and life because you are with your friends and you end up talking, discussing work and relationships, so they are there to talk about. I don't know what it is about holidays. Generally round the pool, we spend hours talking about when you are not on holiday so it is on your mind.
Dan: So it's not really an escape then, is it.

Pamela: Well on holiday, you have more time talking about things. For me to have my friends around, it helps me talk about things. Look at things, do things better, humans are meant to interact and help each other out. I get a lot off my chest on holiday because there is so much leisure time. My friend spoke about this stuff close to her which she didn't mention at home. At home, the conversations are never really in depth.

So instead of escaping it, Pamela and others in my sample, *confront it* – but do so in a way that allows them some objectivity on themselves and it is this which is potentially dangerous territory. The spatial and situational shift in the self allows for some degree of emic reflection. However, there is significant vulnerability in these moments as the false reality of the glitz and glamour of the 'good life' in Ibiza starts to become more and more appealing while life back home looks even more ordinary, even more bland and mundane. It reminds people of what they don't have when they get home. What is there back home for me? Why bother going back to grief? Consequently, home life then appears to them massively out of sync because it does not equate to the 'dreams which they should be pursuing' and have been told are available to them if they follow those dreams. This very often acts as the personal catalyst to seize what they have while they can on holiday and this is a catalyst for excess, *hyperconsumption*, deviance and risk (Chapters 3 and 9). This is another important shift because it allows the self to start to consider the self-promising alternatives which are normally one of two things when they get home:

1. Do something to alleviate the 'shit home life' which may involve some effort, normally pursuing something a bit more and/or changing life direction.
2. Immediately 'get on it' when returning home/come back to Ibiza (or similar places) as soon as possible/consider staying there to work for a season (Chapter 10).

At this point, home responsibilities such as work and family generally intervene for the older group, meaning option two can only be a distant pipe dream. However, some of the younger crowd, who have yet to establish themselves in these areas of social life, are easily swayed to the persuasion of Ibiza and its glitz; after all nothing significant is on the horizon (Standing, 2011), everyone is talking about it, it is the place to *be*, the people who work here seem to have it all, and their personal fantasies can come alive every day (Chapters 4 and 6). It is while the self is astray in this personal no-man's land that the latter becomes a more favourable option for many who then decide to rebook holidays, fail to return on their flight home, or even stay out for the season and/or return next year to start the long process of climbing the casual worker ladder (Chapter 10).

These adjustments to self-identity are important because they lay the foundations for the personal justification for deviant and risky behaviours abroad (Chapter 9). However, there is a further pressure which must be negotiated because everyone around them is doing the same things as they are. While this acts as a rationale to join in, it also prompts for the self to look for ways to supersede the sameness of what exists around them: the same sort of people, wearing the same sort of clothes, doing the same sort of things. On reflection, and with now some misplaced idea of how 'shit' life is supposed to be, the self seeks to negotiate itself out of this position. It seeks to break free of the *unfreedom* which governs its ontology (Chapter 3).

Unfreedom

Thus far, we have seen various shifts in identity as a consequence of the holiday occasion but one feature which does not shift is *habitus* (Bourdieu, 1984) – what they do at home on weekends in the UK which is drink, take drugs, and engage in playful forms of deviance and risk (Chapter 5). If anything, it is these elements of their ontology which remain the same when they come on holiday; they therefore transcend time and space. They come to a place to basically do more of the same as what they do at home – albeit elevated and experimental – but take it to unprecedented levels into *hyperconsumption*. Most have limited interests in seeing the island, capturing any kind of cultural experience and generally doing anything other than getting insanely drunk/getting high. But this is the power of the ideology which is Ibiza's image – one of sun, sea, sex and general hedonism.

The ideology behind (and in front of) Ibiza's image

Tracey: Well, it was a good holiday, despite all the drugs [they don't like 'druggies'], we still had a great time.
Dan: But what was so 'great' about it?
Tracey: [Pauses while looking around at her friends and playing with her hair] Don't know really ... [pause] I suppose all the places were good, the music was good.

Tracey and her friends are not even remotely interested in the house/dance music scene for which Ibiza is famed and yet the only place they were on holiday was San Antonio. They had previously been to Kos, Ayia Napa and Faliraki and, more recently, Ibiza several times (Chapter 6). On their most recent trip to Ibiza, they said the *'reps were like dogs'* and they *'couldn't stand the druggies'* ... but they still had a *'great time'*, didn't they? This is the power of the ideology which surrounds Ibiza and the way in which young Brits have been socially conditioned to 'love Ibiza' and the activities associated with going there such as the booze, the birds, the drugs, the clubs, the brands, etc. They go for the 'names' which are ideologically constructed:

Oliver: People go for the name 'Bora Bora' or 'Amnesia'. The music doesn't interest them. It is the name. They don't know anything about the music but go because it is 'Pacha in Ibiza'. But there are people who come for the music and know what they are talking about. But more and more, people come because of these names.

So because everyone says it is 'the place to be', everyone feels they need to reproduce similar discourses about it to feel included and this is one reason people have to say they like it because if they don't, they look stupid; they are outsiders. It is these thin narratives which nourish the ideological image of Ibiza as the place to *be*. But really some don't know what's going on. Earlier in the chapter, we heard from one of the more advanced and reflective interviewees, Pamela, aged 26. Having been to numerous other destinations, she was exceptionally critical after having had a short break there over the summer of 2012:

Pamela: But that's it. The excitement and the enjoyment that this holiday is going to give you and it keeps people going because even when they don't have it, they pretend they have it because they presume they are meant to have it. Ibiza was like a placebo with lots of people pretending they are having fun but I ask 'why'?

Pamela rightly identifies this ideological blanket which coats Ibiza and goes on to add that it is more about the social envy attached to having done Ibiza as an 'experience' which is prevalent than any deep foundations about why they are actually going to Ibiza. She adds:

Pamela: For some people, it is not about the experience, it's about telling people you have had that experience. When you go on holiday, everyone has this obsession of taking hundreds of pictures: me by the pool, me by the beach, me talking to those guys at the hotel, me in the bikini, that club, this club but you might not have that much of a good time but when you post the pictures on Facebook or whatever they say 'Ah man, what an amazing time'; and they are smiling on their face but that was the only smile of the night. No one wants to come back and say 'Went to Ibiza, it was pretty shit, didn't have that much of a good time, not really into clubs and won't be going again' so you lie and tell people you had a great time ... I don't know if it is the experience or the thought that you are being part of something which millions of young people are and they say they are enjoying it. Because if you don't enjoy it you think 'Everyone is enjoying it, why aren't I enjoying it?' and you think 'What is wrong with me?'

This is the first part of their *unfreedom* – the way in which people subconsciously bow down to the ideological imagery of Ibiza and what to do there; they literally buy into it. If they were truly free, Tracey and her friends probably wouldn't have gone to Ibiza because, although they have been three times, they don't seem to have enjoyed it. Equally, as Pamela points out, having been to similar holiday destinations with similar attractions where similar behaviours take place, she probably wouldn't have chosen Ibiza ... but somehow did. It was subconsciously embedded (Chapters 3 and 5). This is the power of the ideology which is sustaining Ibiza's image and it is embedded in the psyche of many of the people in this book. Related to this is the way in which this ideological image reinforces *habitus* (Bourdieu, 1984) – and this is marked by these young Brits' perceptions of what they should be doing when they go to places like Ibiza (Chapter 5). They think they 'can do what they want to do' but they can't because this represents a lack of liberation, an *unfreedom*: a capacity to only enjoy what they have come to know what they *should* be doing with their leisure time and engage in behaviours that are *expected* of them (Chapters 5 and 6). So in a resort where everyone is doing the same thing, wearing the same clothes, escaping the same/similar problems at home, and uploading similar sorts of pictures on Facebook, how does one break free from this culture of sameness? I want to suggest this is either by seeking 'cultural experiences' and gaining some 'holiday capital' or engaging in behaviour which pushes conventional boundaries of home deviance and risk, and for this cohort, it is the latter which is far more prevalent. This is how this pattern starts – on the first night.

'Getting on it' and 'getting messy': The first night and destination hyperconsumption

We saw towards the end of Chapter 5 how ambitions for a holiday blowout frequently start at the airport and continue on the plane. This, I showed, was partly buoyed by a social construction of the expected practices gleaned from previous experience as well as a hegemonic influence through popular culture (Chapter 6); that is, 'We are on holiday, let's get pissed, this is what we do and have come to learn should happen'. Assisting with this are the airport pubs, the drink offers on the plane and even the subliminal messages on the back of the plane seat – that is, 'We are on holiday, let's get pissed, this is what is expected of us when we go away and, oh look, we are also directed into these practices'. Such is the determination to make the most of this time that it often means powering through and engaging in *hyperconsumption*. This is normally christened on the first night of the holiday: generally considered to be the 'most messy' night as the excitement of perceived escapism from home life comes to fruition. One principal finding in this research is that it is this night which is the most risk-oriented as many young Brits in this sample go completely overboard on their alcohol/drug consumption. Here are three first-night examples based on my field notes in and around San Antonio.

Where's my hotel room?

We arrive back in the hotel; it is nearly daybreak. After some shouting outside between some prostitutes and potential clients, we hear a loud banging at the door down the corridor. After several minutes of banging, we all tiptoe out to peer around the corner at what is going on. At first, I see a tall woman frantically knocking at the door. However, when we walk around the corner, we see another woman, similarly dressed in high heels and dress, lying flat out on the floor. 'Oh shit, what happened?' I say as I step up a pace towards them. The woman who is standing asks me to wake her friend on the floor with her leg collapsed over the other and one hand bent about her head with the other fully stretched out; it is the picture of a car crash as she lies in a heap of arms and legs. There is a lucid smell of alcohol on her breath. I suggest we turn her into the recovery position and we all start to roll over her docile, tall figure. With that, she starts to come around. Within 30 seconds, she has managed to get herself up the floor, and using the wall to keep her on her feet, she says *'I'm grand'* but clearly she is not *'grand'*.

We retreat to the room but the knocking periodically continues. After five more minutes, we leave the room again. This time the same woman who was lying on her back is lying face down on her front and it seems like her friend has abandoned her. Again I tap her and try to establish she is OK but there is no reply. In the distance, I hear the light pitter-patter of trainer-to-floor steps complemented by louder high-heel-to-floor steps; it is the night porter leading the worried friend with the keys to the room in his hand. He tells us not to worry as if this is common and tries to find the right key to enter. He fumbles as if he is being timed but the pressure consumes him to the point that he keeps selecting the wrong key. As he finds it, the other woman tries to reinvigorate her friend with the offer of the water we have given her and there is a groaning response. We again retire to the room. Outside, a group of young men walk down the street eating kebabs, one stopping to urinate on a grey Citroën. Luckily for him, he finishes just in time as two local policemen come around the corner and drive up the same road.

Around 11 a.m. the next day, I overhear the women's voices but from the other side of the hotel corridor. Last night, they were knocking on the wrong door; it was someone else's room. In any case, the woman who was lying down now seems to have made a miraculous recovery and stands reasonably sane in her bikini. *'I remember you'*, she says to me, *'you helped me'* and then thanks me. When I talk to the hotel staff they now suggest that these things don't generally happen because it is mainly families who stay here; perhaps they too are worried about tarnishing their image. However, the reception woman says they only arrived in Ibiza yesterday and they arrived drunk and went out again; seems it was all too much on the first night. *'Siempre los británicos hacen fiesta demasiado en el primer día'* says the receptionist (the British always party hard too much on the first day of their holidays). [Field notes]

Where's my hotel?

It is just after 1 a.m. As we walk toward the drinking strip from some backstreets, we are stopped by a sweating, drunk man with his shirt half open. He slurs at us 'D'ya know where Central Hotel is, pal?' and we shrug our shoulders and look at each other blankly. 'Is that where you're staying?' I ask. And he slurs 'Hotel Central this way?' as if he thinks we are foreign; perhaps because we are not wasted. 'Is it your first night?' I ask and he nods. He then glances in some sort of drunken disorientation up the road from which he has just walked and then down the one he will likely continue. He asks *'this way?'* and points and we say we don't know once again. In the end something sensible comes out and he says 'This way to the taxis?' and we redirect him a little. He stumbles off in a direction which we didn't recommend. [Field notes]

I know the name of my hotel but what's my name?

We stand outside a take-away near el huevo (the roundabout); it is well into the early hours as we notice, lying on the floor, a man half asleep/drunk nearby. The take-away workers on a fag break while mopping their brows, shrug their shoulders. One says 'His friends left him, probably his first night'. I look around at all the partying and fun which seems to be going on then at this collapsed figure wondering how it could have happened that his 'friends' would leave him in such a state. When we approach him, we get a better look of his condition: drunk to unconscious. He lies in a heap with his head against the door and his body crumpled thereafter on the pavement; it is as if he had fallen eight storeys and landed in a pile. When we try to wake him, he says he remembers his hotel but can't remember his name, and returns his head to the awkward door-resting position. We call Ibiza 24/7 just in case. [Field notes]

Therefore, from what I can see the first night tends to be the most risky in terms of the level of excess in which people engage but also the risks external to them that they expose themselves to as a consequence of that excess (Chapter 9). For the jet-setting tourists – long weekenders or those coming for the opening/closing parties in June or September – coming in for a few days, the level of *hyperconsumption* continues because the visit is short and as much needs to be made of the event before the impending trip home. However, for some on holiday for periods beyond a week, likely on a package holiday, there is normally a slight lull in the level of excessive consumption because many cannot keep up with the pace. However, almost all who stay for a week or longer will likely seek to increase their consumption as the end of the holiday beckons; the last night is also a landmark for *hyperconsumption* to risky levels because it symbolises the ritualistic parting (Chapter 10). This doesn't mean the pressure disappears, nor are there any noticeable differences between the male and female groups. Those who 'cannot keep up' are frowned on and, at times, excluded for their weakness to wring as much from the experience as possible. Take this mixed group of friends in their early 20s, who were interviewed on Bora Bora beach in Platja d'en Bossa as they lay over each other drinking beers as if they were in a heavenly paradise:

> I arrive at Bora Bora beach. On one king-size bed are two groups of young Brits from different parts of the UK: one female group who came last year and a male group who are staying in San Antonio for the first time and had heard about Bora Bora. At home they confine drinking to weekends and two relate the experience of being away on holiday to being let *'off the leash'* and their holiday daily routine is one which consists of cider and beer throughout the afternoon, moving onto spirits as they change and shower ('pre-drinking') which entails some games and competition so that when they get to the clubs they don't have to spend much but may buy a few drinks in the clubs. Indeed, one says their money will go on 'drink, drugs, clubs and taxis and nothing else'. As they drink in the sun as if they are kings of the Earth, throughout the interview, they make clichéd references to the way in which one should party in Ibiza: *'eating's cheating'* says one young woman while another says 'It's like we're here for a good time, not a long time'. [Field notes]

These young people have come to believe that this is 'true freedom'. People say they feel 'free' and can do what they want to do in Ibiza but I am asking what 'liberation' and 'freedom' comes from going to a place only to elevate what is done at home (drink more, take more drugs, eat more cooked breakfasts, shop more)? This doesn't sound like 'free will' or 'freedom'; completely the opposite, it is a form of *unfreedom* because they have been bound by a *habitus* moulded by marketing campaigns, adverts, and a significant amount of time in the NTE back home (Chapters 3 and 5). It may appear as a 'free will' decision but this is the ideology at work (Žižek, 2011). This next excerpt gives you some idea about how people are bound to these practices in these places; it highlights how liberty is restricted to doing what they have come to learn they should be doing (Chapters 5 and 6) in a place which is strategically designed for it to be done (Chapter 8). These young women from Liverpool and Newcastle, who get free drinks from the bar owners by going in to 'dance around' (to coerce the male punters), didn't know what to do after a few nights of partying in San Antonio:

Dan: But what is the point in getting fucked [drunk]?
Alice: We came here for a couple of weeks and last night, I didn't want to get pissed. So we just walked around and had nothing

	to do, there was nothing else to do. So we ended up getting pissed for nothing. There is no places to go, no art, nothing.
Dan:	Is it designed for you to get pissed?
Alice:	Yeah but I like that, fucking right I like that.
Dan:	I see.
Alice:	I have been here for three weeks and have been pissed every single night.
Sophie:	When I get home I will be dead.

The resort seems to be in this case the be-all and end-all of the holiday. Beyond there is nothing else, nothing of interest. And when there is nothing else, the only thing to do is to 'get on it' again. This is *unfreedom* for if it was any form of liberty, it would make for more concerted efforts to get out of the resort to learn about the island, its history and other sites of interest. This *unfreedom* – the inability to break free from *hyperconsumption* as a mode of *being* on holiday – I want to suggest can be negotiated in two ways. The first is by attempting to take in cultural experiences. However, for this cohort, this doesn't match with their *habitus* which has, to some degree, come to confine them to the familiarity of cultural practices of the resort (drinking in the bars and/or taking drugs each night). This leaves the almost inevitable second which is by transgressing *unfreedom* into extreme deviant and risky behaviours which are often taken to another level in an effort to break through this culture of sameness framework while at the same time generate group kudos. The actions therefore often appear as more bizarre, more extreme and, in some cases, more dangerous. The former is explored in the next section before the latter is thereafter examined.

In search of the 'cultural experience' and the quest for 'holiday capital'
Here Keely describes a trip out of the resort on a holiday in Bulgaria.

Keely:	We took a trip out for the day, me and the girls. They say to you 'do not go out of the resort', it's one of those places you do not go out of the resort as it can be quite dangerous. But we decided to look for a shopping centre which probably didn't even exist, like you do [laughs], we wanted to go on an adventure. We got lost and it was getting dark and there was signs, I don't know what they were but there was loads of pictures of people's faces, I don't know if it was missing people's faces, I don't know what they were but they was everywhere on the walls, different faces. I can't read Bulgarian so I don't know but there was beggars constantly

Dan: coming up to you, old ladies constantly begging. It was sad to see but you're not going to get your purse out in a place like that. And there was all these cobble streets. It was like something from a movie, it was really quite scary. We finally found our way back. We was on a little bus. It was like something you see on TV.
Dan: Serves you right for your little adventure.
Keely: I know, but do you know what, it was such a good experience [laughs] cos I see a bit of the cultural side. I suppose in ways it's good to say that I've seen that and got through it all in one piece.

Keely has been to Ibiza three times now, having worked in Majorca and been on holiday to Sunny Beach in Bulgaria. What she thinks she is describing here is a 'cultural experience' – something totally detached from what she should be doing – which is spending time on the beach, drinking in the evenings and going out at night.[2] Of course, although in resorts like San Antonio every day and night is different it is often the same in that people get up, recover from their night out, go to the pool/beach where they likely drink more alcohol, eat a bit, drink, get ready to go out, drink, go out and drink, perhaps take drugs, etc.) and, in some ways, they can never know what *exactly* will happen but, in a way, they can know it *will* happen. And the next day is spent recovering and reconstructing the night from digital camera photos for a few hours before they are back on it. Where then is the time to see the rest of the island since becomes the established routine for most.

After exhausting some of these destinations, including Ibiza, 'holiday capital' – that is, seeing the sites outside the resort, perhaps learning the language but at least seeing and experiencing the culture outside the resort – may be retained over a period of time for only a few. These people think that there should be a more meaningful way of spending their holiday time apart from getting wankered and likely going home in debt. Somewhat tired with the routine of excessive consumption, this small group may graduate to city breaks and even to 'travelling', as they say. However, even then, they may not quite be able to evade their *habitus*: the familiar practices into which they have been socialised at home and what they have come to know they *should* do when they go abroad. These girls from Newcastle seem uninterested in anything outside the resort:

Dan: Apart from the resort, are you interested in anything else?
Valerie: NO! Why would you? That stuff is for boring people who go to Sharm el-Sheikh [a package holiday resort in Egypt] and that sightseeing.

Dan: Cultural experiences?
Jane: Yeah, last year we saw the sunset.
Valerie: And we went to Palma [capital of Majorca] to see the cathedral but we were a little drunk.

'Cultural experiences' are seeing the sunset and the cathedral (a bit drunk!) but the truth is other things outside the resort, and outside what they should be doing which is drinking and partying, are not on the holiday radar. Even a return to the old holiday destinations and the old familiar holiday lifestyles may beckon. Take Lee and Tom, for example. After a *'lads' holiday'* in Magaluf in 2009, and after their first year at university in 2011, they decided to go *'travelling'*. Having *'saved £2,000'*, which was their student loan, their excursion was to last six weeks, taking in the sites of Europe. However, they didn't get out of France, and returned broke after only ten days. In 2012, they decided to 'live the dream' and become casual workers in Ayia Napa.

Similarly, we met Liam and Graham in Chapter 5 and saw what they had to say about *'other holiday'* destinations they went to when they were younger. Liam and his friend Graham are in their mid-20s, and they distance themselves from the West End and agree that San Antonio is not the *'proper Ibiza'*. Having been to Ibiza three times in the last year, they say they reserve their trips primarily either for the opening or closing parties with a holiday between (in which they end up going to the drinking strip to get *'headfucked'*). On the last two occasions, they have stayed in Platja d'en Bossa yet miss the *'banter with the other Brits at breakfast'*. In the more international resort, it's *'not the same'* with the Italians and Germans in the hotel – clearly they miss the constructed sense of solidarity of San Antonio (Billig, 1995). Here, Liam tries to further distance himself from the madness of San Antonio by suggesting he is 'cultural':

Liam: Like I am quite cultural to be honest. I went to Ibiza Town for the day and it was fucking lovely up there, you had the castle and everything. He ain't cultural [prods Graham].
Graham: [Laughs] No *you're* not [cultural] ... Just cos you stayed in a hotel where there were loads of Germans and Spanish and you slept there until like 5 p.m.
Liam: Yeah but I still went to the castle and read every single plaque. Such a view from up there.
Graham: Yeah but you went when you were fucked on pills at 4 a.m. in the morning.

Dan: So did you go anywhere else on the island?
[Pause]
Liam: Er, my parents went to the north of the island.

Liam thinks he is 'cultural' because he did something other than what is expected of him and what he expects of himself (even though he was pilled up!). For the few who are looking for some sort of 'holiday capital' – some experience beyond the resort of what they have been socialised into doing – it doesn't work out because many have come to learn that when they are away from home on holiday, they have limited capacity to do anything else apart from engage in unbridled excessive consumption. We return to Pamela who had fallen out with her boyfriend when she went to Ibiza (precisely because she was going to Ibiza). She had first heard of Ibiza when she was a teenager but went to Malia in 2007, when she was 21, with four friends and was *'out every night like a trooper'* on a Club 18-30 package holiday:

Pamela: It was drink as much as you can and sleep with as many people as you can and it was all piled upon you, all these excursions with drink, vodka and ridiculous drinking and kissing games. I wouldn't do an 18-30 again because they march you around but it was the first holiday with my friends and it was good to have someone say 'This is what you need to do, do it' otherwise we would have been a bit lost, 'How does this work and what is the normal routine?' We spent an extortionate amount of money for beach parties, boat trips, but it was good I suppose.

The year after, in 2008, she went to Ayia Napa with eight friends where she was out every night 'partying', and then in Magaluf in 2009 with five friends, although this was not so enjoyable because of the disputes. By the time she booked Ibiza, she knew what she was in for:

Pamela: I went to Ibiza to celebrate graduation and get drunk and have crazy hangovers. Hmm, I'm not sure why I went, it will be the last raving holiday I go on. I didn't enjoy it much. It was clouded by my partner, because it was at the expense of my relationship. I didn't let my hair down as much because I didn't want to flirt or be with anyone because I didn't want to become the person my ex-boyfriend thought I was.
Dan: You say the motivation was to celebrate your graduation.
Pamela: He split up with me because I booked the Ibiza holiday. I think I went to [excitedly and picking up her pace]

	spend time with my friends to celebrate my graduation but [deflated somewhat] it wasn't so fun. At the end of the night, me and my friend were sitting on the steps of the hotel chain smoking and we were just bitching 'bloody meat market in there', 'not doing this next year, need to get some culture from Thailand'.
Dan:	Why? Why separate yourself from them? What is different about you?
Pamela:	We weren't there to have meaningless relationships, get so drunk, take illegal drugs. We had structured enjoyment whereas theirs was uncontrollable, young excessive … over the top, extreme consumption of drugs, alcohol and sex. This was the difference. We were a bit … unimpressed.
Dan:	But you should have had a clue you would feel like this. You've been on these holidays.
Pamela:	Yeah, because I felt like this three years ago in Magaluf.
Dan:	So why Ibiza?
Pamela:	I don't know. It is the most mature place of all of the places. We hoped there would have been an older, sensible crowd there but on reflection we should have gone Prague or Amsterdam, a city break. I don't know why I chose it. It was a poor choice. I think it was subconscious and wanting to like it, but why is it rubbish for me? I wish I hadn't gone. It took me four days to recover and I lost a relationship.

Pamela had matured out of these holidays three years ago, yet still the image and reputation of Ibiza was enough to secure her holiday there. Ibiza was the place where people her age (mid-20s) go and should go to do the same thing they have done in other places. Perhaps she is now realising that her next break should not involve what she has done all her holidaying life; perhaps she may achieve 'holiday capital'. For most others, if not all, this kind of cultural experience doesn't exist and many, I want to argue, find other ways to break their *unfreedom*, to look different. They engage in bizarre and extreme forms of deviance and daring feats of risk which helps them claim back their individuality as well as reap social credibility (Chapter 9).

Transgressing unfreedom

> *'I wake up everyday it's a daydream, everythin' in my life ain't what it seems, I wake up just to go back to sleep, I act real shallow but I'm in too deep'* Dizzee Rascal – 'Bonkers'

What I think the British rapper Dizzee Rascal is accurately describing here is a materialistic mundanity of contemporary Western social life; a life which is not real; a life which is governed by surface-like interactions but at the same time hides the complexities of the human subject (Chapter 3). At home in the UK, on nights out, this takes place as a means to pass the time but these weekend nights out are now constructed to some extent as banal – they have become part of the film of expected practices. Why would so many describe home life as boring if it did not include what goes on at weekends? However, the market has now cleverly capitalised on new forums for excessive consumption such as gigs, festivals and, in the context of my work, the holiday. It is by doing these things, at these venues in these places on these occasions, that people feel liberated; they *feel* free, but they are still participating in the unreal, the materially constructed.

Who, if anyone, realises how much they are bound to a consumer culture in which so many participate? It is this materialistic culture, from which I have shown earlier (Chapter 3), people seek to rise above – to transgress. Existing in the unreal (material life) only prompts a quest for the construction of the real (transgression) and it is from this which comes a pleasurable ontology. So coming abroad into an environment which looks the same back home and caters for such adventures only prompts new experimentation. Because excessive consumption often happens week in, week out back home, when abroad it needs to be taken to new levels; it needs an edge, it needs to exceed all known boundaries. This means on holiday it can appear extreme, almost unthinkable. Look at what one friend of the Southside Crew's gang in Ibiza Rocks got up to on one night. Such was the moment of the holiday, the pressure of the group to do something different and stupid and, as I am suggesting, the increasing need to go beyond conventional boundaries of behaviour that this happened:

Paulie: My mate was arrested last night for shitting off a balcony [one of the 14 others staying with them].
Dan: You're kidding. Shitting off a balcony? What the hell is that? [They all laugh]
Paulie: I know, fucking filthy isn't it?
Dan: I cannot think of anything more raw.

Back home, Paulie's friend would never have done this. Had he, he would have likely been socially ostracised. However, on holiday, it has new meaning. It is symbolic of a time to construct folklore and be remembered. It is a time to rise above establishment, law, social norms, material life – every expectation – it is time to take things to the extreme. Because with the advent of the holiday, there is an extension of this sameness; what they know is everywhere. So how seek difference? How reclaim one's individuality? Engage in behaviour which pushes conventional boundaries into the bizarre and extreme. In summary, I am saying that the holiday (as time) underlines normality of home life and the resort (as space) potentially recreates a simulation of home life (Billig, 1995) – it is unreal, materialistically constructed. The 'cultural experiences' are not on the radar, and thus 'holiday capital' is some distance away and may never be a destination arrived which leaves only the adult playground of San Antonio, the Superclubs and/ or various beach clubs in which to make legendary memories. However, these memories come at a price as it is these spaces which are strategically designed to exploit the general holiday attitudes of these young people as well as cater for their subjective fetishes, and this will be the subject for discussion in the next chapter.

Conclusion

The chapter shows how the hegemonic ideology around Ibiza subsumes different groups of the sample; young or old, with or without responsibilities, in and/or out of work. Many aspects of how their *habitus* has been moulded (Chapter 5); what to expect on holiday, what to do and how to behave (Chapter 6) comes to life. However, this too has some ideological origin because it tells the younger, inexperienced crowd that Ibiza is where they *should* be and this is what they *should* be doing there in their 'youth' while it equally applies to the older group, who are more likely to have work/family responsibilities, that they need to seize this reality before returning to pressurised lives. In Ibiza, *time folds* as the two groups converge to do the same sort of thing – only at different points in their lives – engage in *hyperconsumption*, bizarre and extreme behaviours with the aim of creating as many precious moments as possible which can be reflexively revisited when back home. In addition, they say they escape home life but for most it ends up being an unpleasant confrontation with home reality. By comparison, home appears out of kilter in comparison to the 'good life' they are experiencing. For many, this contemplation is enough for them to draw artificial conclusions that life at home is really shit and that 'living the dream' in Ibiza has all the answers for them. Engaging in these behaviours is also assisted by fact that they are in an anonymous environment which does not reflect immediately the gravity of their actions and so without the home laws and social expectations, a *new permissiveness* is constructed. They say they feel 'liberated' and 'free' but end up displaying exaggerated home behaviours and are at the mercy of the commercial powers of the resort. I have argued they are *unfree* because the pursuit of cultural experiences doesn't seem to match their *habitus* and instead a default setting of excessive consumption is initiated into the realms of the bizarre and extreme. However, capitalising on the attitudes of consuming as much as possible such as drugs, alcohol, food, and unfortunately women, is the role of the resort in facilitating the behaviours (Chapter 8).

8
The Political Economy: Consumerism and the Commodification of Everything

Dan: So you two have been here before [Jay and Nathan]?
Nathan: Same hotel, same place, San Antonio!
Dan: That's interesting because some people are saying that Ibiza is the last one, the ultimate and that other holidays facilitate a step towards Ibiza. But there are people that come back here, year after year.
Jay: It's the name.
Nathan: Mate, I will always come back to Ibiza every year. That's how much I love it. The clubs, yeah? In Southside, yeah, it's called the Kingdom. All the best DJs from Radio 1 come down. Drum and bass in one room and Van Dyke and all that and it is exactly like Ibiza, but Ibiza is like thousands of people but like Southside is like one thousand.
Dan: So it is like the ultimate.
Nathan: It is a clubber's paradise. If you love the trance, house, I fucking love it.
Paulie: I hate it.
Nathan: Yeah, he doesn't like it [to Paulie], don't know about them two [Marky and Jay].
Marky: R&B.
Jay: Yeah R&B.
Paulie: All day long.
Nathan: Any day, mate. Me and him [to Jay as if to secure some common ground].
Dan: What do you do if you are not into the clubs then?
Paulie: I would rather stay by the pool and go in the dingy clubs to be honest with you.

Jay: But we are mates and we'll have a laugh anyway [as if to try and unite group on separate interests].

Introduction

The tensions in individual holiday agendas are important here for they come to fruition in the next chapter when the Southside Crew arrive on the West End. However, what I want to draw your attention to here is the fact that Ibiza now exists beyond its underground dance/house music reputation, which, by the way, is now more mainstream than ever. Those days are gone. As I have shown, some British tourists can't quite locate the reasons for selecting Ibiza for their holiday; some say things like because of 'its name' or because it's the place to *be* but can offer little other reason. While there may be some allegiance to the music, the Superclubs or Ibiza Town shopping boutiques, more than ever people feel they *should* go, to say they have gone, and feel they need to do so as part of life's experience. How is it then that they have come to believe this? Well, over the last ten years, Ibiza's strategic immersion in the media, popular culture, and celebrity stories (Chapters 1 and 5) has been coupled with its rampant marketisation. This has also been married with a commercial transformation of its resorts and Superclubs, and the development of exclusive hotels and beach clubs. However, the island's infrastructure – health, criminal justice, and in some respects tourism – is woefully short of supporting this development. So where the formal economy falls short, in steps the informal equivalent (drug dealers, illegal taxis, etc.), and the British tourists and casual workers have their role to play in this respect.

These advances and changes have occurred at the same time as Ibiza's tourist numbers have started to dwindle; in part, because of its image of 'drugs and crime' but also because of the increased competition it now has with other emerging tourist destinations which offer the same sort of thing (music, clubbing, sun, sea, etc.). With tourist numbers down and fighting an image problem, the corporations and marketing entrepreneurs, local authority, the Superclubs and, perhaps more reluctantly, many of the local community who work in the tourist industry on the island, more than ever, face a dilemma: they then must ensure that the visitors who come maximise their spending in the short time they are there. This chapter explores how this has occurred with a focus on the Superclubs, the West End, private hotels/beach clubs such as Ibiza Rocks and Ushuaia, and the development of booze cruises. The changes to these elements, I want to argue, are reflected in a *spatial commodification of status stratification*: that is, how space and social status become intrinsic to spending power. It is these elements, I suggest, which also help to propel British tourists into *hyperconsumption* and to feel they need to return the year after to attain an ideologically higher status.

The island's infrastructure

A free market in Ibiza makes it difficult to regulate and bring to account private and foreign entrepreneurs who have established large-scale tourist business ventures like Ocean Beach Club Ibiza, Blue Marlin, Ushuaia or Ibiza Rocks. These endeavours have only business interests and look only to secure as much money as possible from the British tourists in Ibiza. While we saw in Chapter 4 that the tourist numbers are decreasing, Ibiza will unlikely fold; it is unlikely to go bankrupt but the threat of profit may be jeopardised unless measures are taken to ensure that the tourists who visit spend as much as possible without noticing too much how much they have spent. This has come, I think, through the strategic marketing and advertising associated with all formats of consumer life and it is this which produces an ideological impetus which is found deeply embedded in the perceptions of the British tourists in this book. It is:

- I feel I need to spend more money to get/be part of this 'experience'.

There were glimpses of these attitudes in Chapters 6 and 7 but to make this work more efficiently in the context of profit, there also needs to be spatial containment in the resorts, hotels, beach clubs or Superclubs, so the ideological impetus becomes:

- I feel I need to spend more money to get/be part of this 'experience' but luckily for me I have all the 'experiences' on my doorstep so I don't need to go anywhere/do anything else.

The way the global corporations, marketing experts, popular media and tourist companies have ingrained this attitude was partly discussed in Chapter 5 but I want to suggest that it is these ingredients which also propel people into *hyperconsumption* and, as a consequence, actions to the detriment of their health and wellbeing. Aside from the impact on Spanish residents and local tourism firms, the health services and criminal justice system must also absorb the consequences when things go wrong.

Spanish residents and local tourism firms

In Chapter 4, I charted how, over the last 30 years, we had seen a massive change in the tourism landscape across San Antonio. The main area to which I drew attention was how the advent of mass tourism, the increasing prevalence of bars and clubs, and the withdrawal of other tourist groups from San Antonio, had started to strangle local businesses because they were almost forced to join the bandwagon and provide services/business to the British unless they wanted to go bust. This process has been further aggravated by the increasing presence of British private firms/consortiums, which have developed new business ventures in the wake of other tourists from other countries retreating from San Antonio and now capitalise on the increasing zonal containment of the British tourists. The most obvious transformation has been the introduction of hotel chains offering 'all inclusive' and the construction and subsequent branding of places like Ibiza Rocks in 2005, and later in 2011 Ushuaia in Platja d'en Bossa. Understandably, the local population are angry because the taxes come from their pocket and more business goes to the British proprietors and the global chains:

Pablo: All the bars on the West End are run by British and they don't pay taxes like me [thumps hand on desk]. And the people who work in the bars are not part of the social security system, they just come here on holiday and stay. But this is illegal but it doesn't matter to the authorities as long as they earn money from it, take a slice of the cake ... People, as in residents, protest but they don't do anything more. The government doesn't intervene it seems in the bars and clubs. They allow them to continue like this.

The result is that very little is reinvested into the island's infrastructure. Indeed, as we saw in Chapter 4, the tourist landscape of San Antonio has changed dramatically and negatively impacted local businesses and this was even evident during the fieldwork period from 2010 to 2012. Two examples of this are the expansion of the fairground entertainment on the beachfront and the rapid construction of Ocean Beach Club Ibiza: an 'exclusive beach club' on the Bay area. In 2010, the former was made up of just a bungee jump slingshot with a burger van next to it; however, by the summer of 2012, it had expanded to bumper cars, carousel and other fairground rides. In this short conversation with a PR girl in her first season in Ibiza for the 'experience' (Chapter 6), it seems that business comes later when the punters drift out of the West End, drunk and/or high on drugs – it comes from the spending on excess provided by the Brits:

Dan: How much is the slingshot [bungee jump]?
PR girl (PRG): €25 each.
Dan: €25! That seems more expensive than it was last year. And to get it filmed?
PRG: [Thinking she can do a deal, perhaps thinking I am drunk] Another €10. Yeah but if there is two of you, yeah? It is €50 and you only pay €5 for the DVD each so it is €60, €30 each.
Dan: Er ... but that's still the same price.
PRG: Yeah, suppose.
Dan: Who mainly does it?
PRG: Like mainly stag parties, like they force them, the stag goes on it. It gets busier later because everyone comes out drunk and they are like 'Yeah, let's do it!'

It's not even a better deal with the discount but it would certainly sound like a good deal to a bunch of drunk young men or women who are away on holiday and have to try and do as many 'crazy things' as possible while they are there. This is one expansion. The other is the conversion of a dirt patch halfway up the Bay area into a plush beach club. Owned by Gary Lineker's[1] brother, Ocean Beach Club Ibiza is a spin-off from the Ocean Beach Club Marbella. Generally open from noon to midnight – remember this because it's important – the beach club seems to be some sort of exclusive hang-out for people with lots of money. It is made up of an enclosed area with tinted, transparent windows (so everyone can look on in envy), fake plastic grass, DJ decks, restaurant, bars and stylish pool and minimalistic, white furniture. At the time of writing, to hire a sunbed, just a few centimetres away from the next, cost €20 plus the ticket entrance price which depended on the 'event day'. Punters can then select which level of VIP they want but unfortunately what comes with this are minimum spends between the 'group': a VIP Cabana or Pool VIP comes with a minimum group spend which ranges from €600 to €1,000 (prices for event day and special event day respectively); VIP Beds range from €750 to €1,250; or if you are feeling really, really special and self-indulgent, the Owner's Bed ranges from €2,000 to €5,000. And this is just the accommodation. Food and drink come on top and it ain't cheap either (plain burgers are upwards of €14).

The social company could not boast more social kudos: celebrities and Premier League footballers. On 23 July 2012, *The Sun*[2] reported that Mario Balotelli, a Manchester City football player, hired a VIP table and spent over £4,000. But he has not been the only one for Joe Hart, Andy Carroll, Michael Carrick and Danny Welbeck, all of whom earn thousands of pounds a week playing football in the Premier League, have partied there so the people in my cohort don't mind paying through the nose if it means they too can appear as one of the elite, one of the celebrities:

Graham: Ocean Club Ibiza.
Liam: And it's a copy of Ocean Club Marbella. Bit more classy, bit more upmarket. During the day it's like pool parties.
Dan: And it's €50 to get in, isn't it?
Graham: Don't know how much we paid to get in. We got a bed there, our whole bill was €1,600.
Dan: €1,600!
Graham: Yeah there was eight of us. So €200 each ...
Dan: ... in a day, fucking hell.
Graham: We got a four-and-half litre bottle of vodka for €950. Then Tulisa [Contostavlos, from *X Factor*] turned up.

As I was saying, the campaign for their money is everywhere and when money is increasingly diverted into these businesses and with the increasing prevalence of the global chains, the local tourist businesses suffer most, confirmed in this conversation between two local businessmen working in the tourism sector:

Alberto: In my shop, I sell things which do not interest the British. They don't buy from me. In five years, this shop will not be here. I will have to sell.
Tiago: Same for me and my restaurant. They go to Burger King, Pizza Hut, places like this. It's like we don't exist.
Alberto: Without the intervention of the British government, these tour operators and private organisations will have more control, more power, more drugs ... I don't like it that the British government blames it on the Spanish. Almost all of the bars and clubs are rented to British proprietors if they are not already the owners. The Spanish may own the boats which do the parties but it is all organised by the British tour operators and the private organisations. It is not the Spanish.

Spending is increasingly contained (all inclusive/beach hotel/beach club) and this has ramifications for the local tourism businesses (Chapter 4). But when British tourists come and indulge in what they think is infinite hedonism, the consequences of their behaviours also place an immense burden on the health system, which is not only poorly set up to deal with basic problems but also far from equipped to respond meaningfully to reduce the harm they cause themselves and to others around them.

Health

One of the major drains on the island's infrastructure is what it must do to respond to the problems British tourists bring when things go wrong. The main concern is that, for some time now, each year the health system struggles to deal with the number of Brits who check in for all range of problems from cuts and bruises, to leg breaks, to heart attacks and, unfortunately permanently in the morgue. There is evidence of these mishaps all over San Antonio:

> Outside San Antonio hospital, another ambulance rolls in with another young male Brit in the back. As he is wheeled out of the back of the van, he groans and looks at his twisted leg as if it is not his; there are severe cuts and bruises down the side of it. It is a potential leg break, the ambulance workers tell me, sustained when playing drinking games near the pool but made worse by some sort of metal sticking out by the poolside. After he is wheeled in, the staff raise their eyebrows as they receive another call-out to respond to a pool accident. [Field notes]
>
> I leave one bar and it is now early evening and a man limps past, walking barefoot on the pavement; his T-shirt has bloodstains all over it and he has a damaged ear. When I approach him to ask him if he is OK, he says he was beaten up three times and has no money. He drunkenly limps off – we are to meet again later on to learn his name is Mark. [Field notes]

We'll come to Mark in the next chapter. In San Antonio, there is only minimal frontline support for drug and alcohol use and the consequences of injury. Based in the heart of the West End, Ibiza 24/7 – a religious charity which offers a kind of informal harm reduction service – are the only real agency on the island who proactively go out to help people who find themselves in a complete mess from the excessive consumption of drugs/alcohol. Normally, they will walk them back to their hotel or take them to hospital but they also have a wheelchair at their disposal – in case they end up having to rescue people who are borderline unconscious, having pissed and shat themselves.

Although the work Ibiza 24/7 do is extremely commendable, given the volume of problems which occur in this resort alone, it feels as if they are fighting a losing battle. Otherwise personal safekeeping is the 'responsibility' of the tourist and this is evident in all the rhetoric of the Foreign and Commonwealth Office (FCO). In 2009, the FCO launched a 'stay safe' campaign in an attempt to clean up the Brits abroad image by giving out leaflets, cards and postcards with messages like 'know your limits'. It's just, on holiday, these people tend not to think about 'limits' or even 'limitations' but the reverse: 'exceeding limits' and 'transgressing limitations'. Typically, the messages are loaded with a 'take care of yourself on holiday' discourse. One more recent campaign by the FCO featured an interview with someone who fell off a balcony, in which the message was essentially 'everyone else don't do it, look at me'; it is made to make people think twice before they do, but the people I talked to and spent time with are not really in that mode of thinking when there is nothing but the 'good life' around them and a seemingly infinite opportunity for fun and excess (Chapter 9). This is because the party goes on, no matter what the health cost. Look:

> In the café near el huevo (the roundabout), it is nearly 9 p.m. and the young male and female groups start to drift towards the West End. As I order some chips, bleeding from the knee and arm limps one young man who is trying to catch up with his group, dragging his leg behind him. He can't quite hear the banter in front of him and can't quite get involved but he so obviously wants to. Still he soldiers on in his new Hilfiger T-shirt which has now accumulated patches of blood. In some obvious pain, the blood drips down his arm and onto the floor; one very large splash drops just a foot away from me. In fact as I look behind him, he has left an identifiable trail and then without realising, a young woman in next-to-nothing smudges the blood into the grey pavement with her high heels. [Field notes]

> It is 5 a.m. and I get in a taxi in San Antonio. I strike up a tired conversation with the driver and we discuss the recent death of the young British woman who was run down by someone who was intoxicated with cocaine and alcohol. As we talk about it, and against the darkness which juxtaposes the flashing lights, we pass the area where she died. A mound of flowers hug the surrounding area yet people pass as if the morbid shrine is invisible; the party continues. [Field notes]

It's a similar story for the tourist companies like Thomas Cook and First Choice; on their websites, between the glossy pictures of the destinations and blogs of what it's like to be a club rep, it is difficult to see any form of practical harm reduction advice which may be of use to anyone wanting to know what 'not to do on holiday'.[3] Instead, most of these companies' websites act as blatant promotion for excessive consumption and sell packages such as bar crawls to the tourists on arrival. For example, in 2008, Club 18-30 advertised Faliraki under the headline 'Faliraki. Because you've got the rest of your life to act your age';[4] note the play on 'youth' and 'experience' (Chapter 7), and the general association that going abroad and getting wasted is entirely expected.

Aside from the frontline support of Ibiza 24/7, generally the health infrastructure can deal with potentially minor issues which, in the main, result from drug/alcohol use. Some Brits, however, wanting to avoid holiday costs for injury often brave through their problems until they get home: they wouldn't want money to go on unanticipated costs when it could go on drinking and having a good time. However, for more serious cases of drugs and/or alcohol, the only place on the island which has specialist care is at Can Misses hospital on the outskirts of Ibiza Town. The hospital, which serves as the main medical point for the 90,000 residents as well, has only 110 beds. Local data from 2005 to 2010 shows on average, around 850 tourists per year were admitted to the emergency department for excessive drug and/or alcohol use and that, although hospitalisations started to decrease from 2005 onwards to 519 in 2009, they have since risen sharply in 2010 to 805 (Govern de les Illes Balears, 2011).

Health staff put the reduction down to the enforcement of a new law which prevented 'afterparties' and the increase to the prevalence of 'booze cruise' boat parties and the erection of private hotel/beach club/music venues such as Ibiza Rocks and Ushuaia. In 2011, this figure had climbed again to 1,064. In the same year, emergency admissions were 69 per cent male and 31 per cent female, and the primary age group were between 20 and 30 years old (68 per cent). Ecstasy use has increased considerably from 27 per cent in 2010 of all cases to 41 per cent in 2011 (Govern de les Illes Balears, 2011). Nearly one in four of these hospitalisations in 2011 were British (23 per cent) but this does not tell the whole story though, for many British hospitalisations or even admissions are not recorded and don't come to the attention of the authorities. And here's why.

Despite the construction of a small, new hospital in San Antonio, hidden opposite some back streets close to the bus station, its main service is to the local community. Seldom are British tourists found here for long. Very few I am told know about it and even then only some present a EHIC (European Health Insurance Card) to get free healthcare. However, in San Antonio hospital, if I have it, my EHIC doesn't cover treatment for drugs and/or alcohol. Neither are there specialised facilities to deal with these kinds of problems. An initial assessment which is payable will determine this and a referral, which is also payable, will be made to Can Misses where all treatment is payable. Most Brits, it seems, go to Médico Galeno – the very obvious and very small private clinic (which has even fewer specialist facilities) which sits conveniently between two bar/nightclubs near el huevo on the beachfront of San Antonio. There they are quite happy to treat minor injuries at a premium. But either way they win – without insurance, they charge what they like; with insurance, they charge what they like and they can claim it back and so can the tourist:

> Outside Médico Galeno sit two ambulances ready for a call as I strike up a conversation with the staff. I pretend that I have a friend who is recovering from a night of excessive drug and alcohol use who I think may need some treatment. I concede I am worried about him which is why I have come in to find out about what happens. It matters not whether he has an EHIC card for they will charge. She says for the doctor to call out, it will cost €140 but that he can claim it back if he has travel insurance. When I ask if that is possible given that he has taken drugs and alcohol, she says *'no problem'* because 'normally we leave the reason blank so they can claim it back. If we put drugs and alcohol use, then they would not be able to claim it back'. [Field notes]

So while on one hand, the island's infrastructure struggles to deal with the volume of emergency referrals, on the other, its health system is quite happy to swallow money from tourists since it knows that excessive drug and alcohol use play a major role in its own preservation. As I write in 2012, a new private hospital is being constructed next to Can Misses, perhaps funded by money generated from this process. So one significant method for economic survival, especially in the context of healthcare, is precisely through capitalising on the very same tourism from which Ibiza claims distance. This is symbolic of just how awkwardly embedded this type of tourism is in Ibiza's economy and the same goes for drugs, the police and the criminal justice system.

Drugs, the police and the criminal justice system

Law enforcement and the criminal justice system seem similarly poorly equipped to tackle the 'tourists' behaviour' as it does the kind of tourism which attracts and generates crime, disorder and drugs in San Antonio. In and around the West End, there is a concentration of violence, prostitution, drug dealing, drug use and general disorder. On most nights throughout the summer, one can expect to see these problems, or a combination of them, in and around the streets of the West End. San Antonio has an extremely open drug market with an array of drugs available at varying costs from Nigerian or Senegalese distributors, casual workers, and some British tourists who, in particular, come to the island with specific intentions to deal. Simon, and to some extent other members of the Southside Crew, it seems had some drug connections in Ibiza, but here he corroborates this:

Dan: I read there is a lot of British guys here dealing drugs.
Simon: All they do is come over here for four months of the year, come over here with shitloads of gear [drugs], all plugged up [on or in themselves] cut it so put a load of shit in it, work for four months then go home laughing. They party and sell drugs for four months and go back.
Dan: Do they operate in the clubs or do they get someone to operate for them?
Simon: They get someone to do it for them. See them black geezers [we all look over at four sitting in the shade selling sunglasses, looking bored and smoking weed], right, one guy will be running 50 of them. He'll ask them to spread out across the island, give them top-ups most nights.
Dan: Right.
Simon: Yep, one guy will be running this and he will be making millions ... Come out here, all plugged. All in their bodies. Put it up their bums. I know guys doing that. That's how they get it in. Some people can take kilos, 50 kilos. They call it suicide missions.
Jay: You get like gangs doing it. They are here for business, mate. Say there may be like a group of them, 15 of them, come over from Manchester and just do it. They caught a load of them last year.
Simon: Ibiza would not exist without drugs. That is why Ibiza is so good, people come out here, get high and party.

Jay: These guys [African guys] just walk around all day selling. The police could come around here and nick all them black fellas and they would all have drugs but they don't do it because they need it to keep Ibiza going.

Drug use and drug dealing are therefore embedded in the economic structure of the island's informal economy. Without the informal economy, the formal economy couldn't function which means they exist in a form of symbiosis. On a broader level, drug distribution in San Antonio, it is said, is undertaken by gangs from Manchester and Liverpool – although this study cannot corroborate on the precise nature of these networks. In Platja d'en Bossa, and to some degree, Ibiza Town, the Italian Mafia are said to be in control of drug distribution through established legitimate businesses. This was qualified by a convicted drug trafficker:

Manu: In San Antonio, there is a lot of drugs, all sorts, cocaine, weed, pills, amphetamines, methamphetamine, ketamine, whatever you want. The British are there selling [in San Antonio] but Platja d'en Bossa is full of Italian dealers, you know in Bora Bora, Space, and Ushuaia. Some Spanish also sell but it is mostly the Italian Mafia which control the space around there and in some of the clubs.

The first problem with the response to this seems to be a strategic conflict in policing agendas. At the bottom of the ladder, the *Policía Local* (or local police) of San Antonio, generally have only authority over prohibiting certain behaviours such as drunkenness, minor incidents of disorder, administration and monitoring of casual worker contracts, responding to residents' complaints and the like. Their main, day-to-day difficulty is balancing this type of party tourism with the residents' and local businesses' quality of life and it is this dichotomy that never seems to get resolved: as the residents see it, the more and more behaviours become permissible, the more and more helpless they feel that change is possible. This is mainly because the *Policía Local*[5] neither have the authority nor the resources to deal with the more significant social problems incumbent in the area even though, over the last few years, they have been increasingly drawn into large-scale drug raids; something which they are generally unaccustomed to and unfamiliar with in terms of strategic response. In fact, as I joked one day after an interview with them *'that it won't look good'* while being dropped back at my hotel in a police car, I was then told that the same officers raided my hotel in 2011 and arrested two British tourists for possession of 3,000 ecstasy tablets and €2,000 in cash. The *Policía Local* are increasingly asked to do these things because of the ambiguity of responsibility between the *Guardia Civil* (Civil Guard) and the *Policía Nacional* (National Police), which would normally be expected to respond to this level of drug distribution. Therefore, although only periodic major raids are made by these police bodies, it still leaves a major gap in responding to drug markets and drug use:

Alberto: We have 50 local police, each person having a daily shift of seven hours, and you ask yourself 'What can I do?' They can go around where there is noise and minor things like drinking on the beach but they can't do the large stuff like deal with drugs and/or prostitution, or even start to tackle what is going on [drugs] among the [casual] workers. That is for the *Guardia Civil* and the *Policía Nacional*.

This is further countered by the difficulties in organising the raids in San Antonio. A common feature of the market is that many British drug dealers stay in large hotels, staying in one room for one week before moving hotel, and in some cases, change their names. This makes it difficult for the *Policía Local* to intervene, and even though the *Guardia Civil* make annual raids, and publicise them through media channels and posting them on the internet in an effort to propagandise their success, still the drugs come in, continue to be bought and are consumed. Much of the street dealing takes place through illegal African immigrants from Nigeria and Senegal, British tourists by mobile telephone and the closed circle network of casual workers who play a large role in the street-level availability and promotion of drugs. All of this is further perpetuated by, what is perceived to be by most, a fairly liberal criminal justice system in the context of drugs:

Manu: The police used to catch me all the time selling drugs on Bora Bora, near Ushuaia but they just took the drugs and let me go. I have been stopped dozens of times but only spent one night in a police cell and never spent one night in prison here. Some people make agreements with the policemen to pay them off to turn a blind eye.

Even the *Policía Local* agreed themselves:

Dan: Is this permissiveness to use drugs reflected in the criminal justice system?

Andreas:	Yes. A judge that says 80 pills is for personal use or for a few people, come on!
Juan:	This is obviously more than just personal or even shared consumption and it makes us frustrated, and we ask ourselves 'why' are we doing what we are doing.
Dan:	Of course.
Juan:	All of your work for nothing. Something which my boss says is that to be in the police here is to have tolerance as much as frustration. You see nothing happening, again and again and again.

Thus, when the island and in particular San Antonio swells with tourists, so the infrastructure has to adapt to issues which surface. We have seen how this takes place in the context of the tourism and health sector, as well as the criminal justice system. However, such is the influx of the tourists and the appeal for making money, that the formal economy often is complemented by its informal brother because of the demand the tourists create (drugs, transport, and general modes of excess).

The informal economy

Unsurprisingly the main problem associated with the informal economy surrounds the demand, distribution and sale of drugs. Largely associated with the music and 'experiencing' the nightlife, the demand for drugs draws certain social groups into the web of drug distribution in an effort to either 'get by' or make a substantial profit. The people who are perhaps 'getting by' by selling drugs are the African immigrants who also survive by selling fake branded sunglasses, CDs, wallets and the like. In the main, they hang around in the beach area during the day and the West End during the night with these commodities as a façade to their other business of dealing in drugs.

As Simon has said earlier in the chapter, other British tourists come to Ibiza with the sole intention to deal drugs. For example, in one interview, Jackie, on her first holiday to Ibiza with her friend Becky, stayed in San Antonio. Quite quickly, after a few nights on the West End, they made friends with their hotel neighbours; three guys their sort of age from the UK. However, each night while they were there, another five guys from Essex climbed up the railings to stay there. They weren't tourists or workers but staying only to sell drugs. Buying the Es at €3 and selling them for €7, the five young men, who had already been in Ibiza eight weeks by July 2012, estimated they would leave Ibiza that summer each with between €10,000 and €15,000 profit. The third group involved in this ground-level distribution of drugs are the casual workers, known as PRs. Some arrive to work in Ibiza, and on realising they need a National Insurance number, have to fill in forms and register with the police, instead fall into opportunities in the informal economy. These workers in Ibiza, as we have discussed, are 'living the dream' and are normally complementing their salaries by selling drugs, confirmed by one local police officer:

Andreas: In the drug market, the bottom of everyone is the casual worker. They come here looking for some sort of life direction, they say 'I want to work here' but then go to party. Then they don't get a good wage selling tickets or whatever and someone says 'Apart from this, do you want to sell drugs?' and it is as easy as that.

And:

Oliver: Everyone works in the drug market here. If it's not selling drugs, it is renting out houses or apartments to store drugs or let people stay there who sell or use drugs. The worst are the British who run out of money when they are looking for work or can't keep their jobs and they want to party, they end up selling drugs or in this position. So they have to earn money to live, to survive ... Once you are in that network of workers, people can connect you quite quickly. So if you are in debt or need an extra income, they can just put you in touch. A lot of people come here to work but end up selling drugs. It is easy money because everyone wants drugs. The people who come never wanted to work, it's obvious to me because they so easily drift into the party.

The casual workers say they don't know too much about the drug hierarchy because what they got is through 'a mate who knows a mate'. Such is the demand that these people often, while working selling tickets or persuading people in bars, will at the same time sell to the customers going into those places. Francis, who is now in her late 20s, used to be a casual worker in San Antonio from 2003 to 2007. She said 'As a worker, you know other workers will deal, you know, we had people living with us that had pillowcases full of weed or pills, stuffed in the house.' Take this young man who had come to Ibiza to sell Pukka Up booze cruise tickets but quickly ran into debt. Thankfully the opportunity to sell drugs came about so he could potentially leave with a profit:

> A small army of people pass by me on the Bay area from the marina. In the middle is a tall young man orchestrating them and directing them to buses. On the booze cruises, it is always a 'sell-out' and he says 'Everyone is dancing, getting drunk, playing games, getting on it, partying'. He gives me his number: firstly in case I want to join Pukka Up but secondly in case I need drugs. In the month he has been here, he has spent £2,500, adding that 'It's the way to have fun, with money'. When drugs enter the conversation, he says the *'black guys'* sell duds. Cocaine can depend but a *'pretty good cut'* comes at €60 while Ket is €40, *'maybe €35 a gram'*. He then boasts about his view overlooking the Ibiza 123 Festival and how he has had no sleep because of the partying. *'Now'* he says 'I have a little party and shift 20 pills a night', splitting the profit with his mates. *'If it goes well,'* he says, 'I will go home with £5–6,000 profit'. [Field notes]

A summer in Ibiza, where 'the birds and drugs are on tap' and where 'living the dream' and coming home with money is likely always appealing. Continuing, he says:

> PR man: It's crazy out here. Like people I know, loads of people are doing it. They come out here just to earn money like that [drug dealing]. They are like 'Right, I am going to earn some money'. I know a few lads out here who are doing the same thing out here as me and we are all making profit. Trust me, you want to buy stuff off us ... [boss comes over] so if you want to buy tickets off me, let me know.

In San Antonio, there is also always demand which is why a similar form of operation in the informal economy is that of illegal taxis. While official taxis do their best and race around the island, they are also in short supply as the night draws close. The night-time period is the money-making hour and in the absence of official taxis, other alternatives become available. In some respects, as the *Policía Local* reveal, some of them are connected to the drug market networks – taxi drivers also deal and use drugs. Having taken several taxis myself, I wasn't too surprised that on one journey the driver had his window open when smoking a large spliff. These can be from locals driving around looking for punters as much as the British who hire cars and approach the people at the end of the queue to see if they need transport.

The demand is not exclusive to transport and drugs but also sex (see Sanders, 2005; 2008). While numerous strip clubs line the streets of the West End, at the bottom in their groups of anything from 2–3 to 6–7 linger prostitutes. As discussed earlier, the police can do little for this population who seemed to have arrived in Spain as a consequence of political turmoil in their home countries in Africa. Some ended up in the hotel industry as domestic cleaners but the global crisis of 2008 saw many lose their jobs; they therefore occupy a very precarious position in the resort. Generally only coming out when night has fallen and targeting the more intoxicated British tourists, they mostly work in groups, offering the promise of sex and are said by police and PRs to rob clients by *'hacer pingüino'* (giving a blow job while robbing their pockets and then running off – the penguin reference being the young Brit chasing after them with his trousers around his ankles). Again these women are to be found most nights offering these services, much to the general acceptance of the British tourists who see them as commodities to be consumed in the space. In conversation with these two prostitutes in their early 20s as the party was in full swing on the West End, the level of suffering appears brutal as it seems they are not only victims of political persecution and structural exclusion but also perpetual violence from the police and the British tourists:

Princess: We left our country to look for survival. To live. I have been looking for a normal job for so long, I go to offices, hotels, they tell me no work because of the economic situation. They even letting people off [making them redundant].

Precious: I don't want to steal so it's better to use what we have. I ask for a euro from many of your friends. Also if I go there and steal the shopman beats me and the policemen come and beat me.

Dan: The police beat you?

Precious: YES!

Princess: The police beat me, they come and lock me up.

Princess: Some of them assault you. Some laugh, some we just target. At the end of the day, we will change their minds. Some of them want to kiss but we don't want to, that is the problem. [Another prostitute interrupts us again and if feels like she is trying to pry them away from us]

Dan: What did she say, is she OK? She seemed upset.

Precious: She is saying no work, she complains.
Princess: She also says the British people are crazy people because they insult her saying things like 'Fuck me, suck me, bitch, bitch'.
Dan: The British people said that to her tonight?
Princess: Yes, well we are suffering. I feel tired, my leg is killing me. One man kicked me here [points to the back of her leg] last night. Because I was calling him 'Let's for a good time', he said to me 'fuck off, fuck off' and I say to him 'Who are you?' He slapped me in the head and kicked me here. As soon as I have my money for my hotel, I go home.

They estimate there are 50 other women like them working in the West End alone. They are considered to be robbers by the casual workers but they maintain that they mostly simulate sex to avoid diseases and abuse. These women are the forgotten voices of the resort. The growing problem of this informal economy has slowly but inevitably attracted the attention of the authorities and certain efforts are made to combat it. However, as we will see, as soon as those endeavours solidify, new methods to ensure the party continues are found, leaving formal responses looking limp and redundant. This is because there is an incessant pressure to win visitor spending and this can only take place if tourists are contained and their spending is allowed to disappear into the realms of the ridiculous. The next section examines how these spaces operate.

Concentrated spaces designed for consumption: The Superclubs, 'booze cruises', private hotels/beach clubs and the West End

Travel agent: Well also the problems are the boat trips. People let them on with drugs and alcohol and there is no maintaining them, no regulation. We see the consequences. The other day, a body was found floating in a hotel swimming pool just up the road here [a mere 50m from my hotel], no passport and the authorities are still trying to work out who he is and where he is from. The other week another British woke up having been robbed and raped, woke up on a beach.

This quote is from a short interview with a local travel agent in San Antonio, whose company reluctantly advertises and sells, among the cultural trips available to the island, booze cruises, entrance to the

Superclubs and particular events such as Ibiza 123 and Zoo Project (Chapter 4). The issue I want to draw attention to here is that of regulation – or lack of it. On an island where the campaign for the tourists' money is everywhere, the tourist spaces have to be designed and marketed in such a way to maximise spending. This often means that a blind eye is turned to problems so that the money continues to fill the coffers even if the chances of deviance and risk multiply. Here I briefly discuss how this takes place in the Superclubs, booze cruises, the West End and private hotels/beach clubs like Ibiza Rocks, Ushuaia and Blue Marlin.

The Superclubs

The Superclubs play a substantial role in the way in which consumption practices take place, with many charging large sums of money to get in, to buy drinks or to be part of the VIP area; beers can be as much as €10 and single mixers upwards of €20. The longer the punters stay, the more they can spend. While the young Brits are already happy to part with large sums of money to see their favourite DJ, as we saw earlier in the chapter, they are equally happy to have said they have 'experienced' the club – even if they know nothing about the music or the DJ. However, the night won't last long for many if they decide to spend their money at the bar; a round for five or six costing upwards of €100 at a time. Look at how much this young man spent in a club:

Eddie: £500.
Dan: In a day?
Stevie: Yeah, in Amnesia.
Dan: Certainly is amnesia if you spend that much. Fuck me, that's a lot of money.
Eddie: It's a joke, ain't it.
Dan: Well, I don't know. How do you feel about that?
Eddie: Yeah, well he [points at Steve] thinks it's like monopoly money. So why don't you tell him how much you have spent in a month?
Stevie: Nearly £6,000 [in a month]. Going out, drinking, clubs.

Eddie and Stevie give the impression that money has no value when what is really happening is that they are conspicuously consuming to try and appear higher up the class ladder and seizing the holiday moment (Chapter 6). Not all do this. To avoid total bankruptcy, some pre-drink in the hotel after a supermarket carry-out purchase and/or head down to the West End to get loaded on alcohol at one of the organised 'pre-parties' connected to the clubs which means they do not have to spend as much in clubs. Like Harry and James in Chapter 6, this is not how they would want it as most would want to feel as if they could flitter their money away without a care – and it is the younger crowd, taking the short cut, who tend to do this. The Superclubs are clever though. They know that there will be this less experienced group who are ignorant with how they spend their money as there will be ones with spending power. Either way British tourists are paying to be identified with a brand, they are paying for a lifetime's experience: they are celebrating the 'good life' or/and 'living the dream'. The Superclub bosses also know that people will come and take drugs on the premises which is why some permit certain dealers to operate and deal to punters – only there could be consequences for those who may attempt otherwise:

Manu: But you need permission really to deal drugs in the clubs otherwise they will beat you and throw you out. I remember seeing one guy who had tried to do this himself and I was out the back when they brought him out of Space; four guys, Mafia, came out and beat him until he was unrecognisable; I mean he basically had no face, no teeth, nothing.

So to some degree even spending on drugs is contained in the Superclubs. In the main this is ecstasy, cocaine and ketamine; the important thing to note that generally this is another option for those not wanting to spend much money on alcohol ... but they will need water to keep hydrated. The unfortunate thing is that water costs €10 so this is another victory for the Superclubs. There have even been cases where young people have collapsed in clubs because of dehydration but no one is too bothered as long as money fills the coffers. The Superclubs also have their own merchandise in little shops in the club. If the clientele are not into the music – which increasingly they aren't – they can at least leave early with a T-shirt with which to remember the occasion. Either way, people spend once inside the club. These field notes from a night in Space confirm these points:

As we enter, we pass the Space shop and into the main dancing area, where on my left is the sparsely populated VIP area which has some better-looking seats, a man to guard entry and a private bar. In front are a mass of people who all face the DJ, most with sunglasses on. Their hands move in time to the pounding bass which reverberates through the floor. It easily penetrates me through my feet and it feels like someone has turned up the volume on my heart while at the same time installed a large sound system in my head. The next room has four bars, and as we enter, a retro house sound excites a rather different crowd; once again, the dance floor is rammed.

I go to the toilet where there is a suspiciously long queue stretching out to the dance floor – but not for the urinals ... for the cubicles. In the queue, several men get agitated; in particular one who starts to hit and head-butt the doors starts swearing. Fortunately, I am not taking drugs so don't need to queue there and skip it after realising that the queue for the urinals is almost non-existent. As I take a piss, I start talking to the person next to me who says he is here for the *'booze and clubs'* but drugs don't seem to be on his agenda. We wash our hands, and between the queue and the commotion is a Spanish woman in her 50s handing out small paper towels for people to dry their hands. She looks wearily at the numerous people ignoring her as they come out of the cubicles. I go over to talk to her; it seems to be a welcome break for her to actually have some human contact:

Dan:	[In Spanish] Why is there a long queue for the toilets?
Toilet cleaner (TC):	Everyone knows but does nothing.
Dan:	What?
TC:	Drugs, obviously.
Dan:	I see.
TC:	If I catch them I report them, but it is difficult because I need to see what they are doing and I can't go into the cubicles. [As she raises her hands filled with paper towels held with rubber gloves] Everyone knows but does nothing! It's not going to change.
Dan:	Do you tell security?
TC:	Yes but they don't do much, perhaps throw them out for dealing. It's the people who don't know what they're doing which cause the difficulty [those who have not taken drugs before or who are not experienced drug takers]. But it's not going to change, it's too late.

The first thing to note is the prominence of the VIP suite as you walk in – this is where one day you could be if you came back with the money to be there (Chapter 10). In just one room, there were four bars and there appears to be an open toleration of drug-taking because those people will need to spend money on water as well as the man I talked to at the urinal will need to buy alcoholic drinks. In addition, there seems to be no consistent monitoring of numbers in some of the clubs on particular nights which are popular: this is so the club can cash in regardless of people's safety and wellbeing. Indeed, there have been several cases of people collapsing and dying in the clubs as a result of over-intoxication of drugs/ alcohol. Where council workers and harm reduction staff have tried to intervene, there seems to have been little progress. A law passed five years ago prohibiting the clubs to operate/host 'afterparties' has instead shifted the continued partying to private hotel beach clubs and booze cruises – the State can't intervene, the party continues and risk multiplies.

Booze cruises

Following the passing of the law banning 'afterparties', there was an increase in the prevalence of boat parties otherwise known as 'booze cruises'. Says one officer from the Ibiza Town police, 'They say that afterparties are prohibited but all they did was change the name: they are called beach clubs and boat parties.' Often sold with the promise of pulling the opposite sex (Andrews, 2005), they generally leave from Platja d'en Bossa, Ibiza Town and the marina in San Antonio at varying times of the day. They can range from €30 upwards to about €70, (although some more exclusive ones can exceed €100), can last anything from four hours to all day, playing music, perhaps offering free bar and food. Some like Pukka Up act as pre-parties to the main bars and club nights while others are used for 'afterparties'. Here is one example; this is the WTF Boat Party advertised on Facebook:

> WTF Boat party Setting off from San Antonio at 6pm with DJs Jimmy Lee and Jon Slater (who will also be playing this week for cream @ Amnesia, Carl cox @ Space & Tiesto @ Privilege) taking you through four hours of mayhem partying, three hours into the journey we will watch the sun go down … then back to San Antonio for an after party at Linekers bar where I have buy 1 get 1 free vouchers all night! Price is €45 each or can be pre paid as £40. Included in this price is four hour boat cruise with sangria cocktails welcome drinks. Then afterparty @ Linekers.

Note how the association with the boat party DJs are with those of the Superclubs and there is a real sense of encouragement to 'get on it' because afterwards the punters will be herded into Lineker's Bar to take advantage of the 'buy one, get one free offer'. In this way, as we learnt in Chapter 4, the tourist in this context is not 'free'; they are being steered quite strategically into more excessive consumption in other businesses owned by the British. The expansion of 'booze cruises' has seen a shift in problems of drunkenness and disorder in San Antonio's West End. Now, say the *Policía Local*, people get drunk earlier:

Andreas: On the first day, the tourists arrive and go on the beach. A man appears to sell them a boat party ticket at 4 p.m. the next day but you have to turn up at 3 p.m. to claim the ticket and pay a deposit. [Gets out about 20 different boat party tickets] So for example, €40 for a boat party leaving tomorrow at 4.30 p.m. all included and you have to be there the next day at 3 p.m. to pick up the wristband. I pay €10 deposit. When I get the wristband, I get two free shots. Then I get on the boat where all food and drink is included until about 8 p.m., extremely drunk. Then by 11 p.m., I am so drunk, I don't know what's going on. So I was drinking from 3 p.m., much earlier than I would have been had I not been on the boat party. A few years ago, we used to see the main problems on the West End at 2–3 a.m. but now it is 11 p.m. because people are drinking earlier.

This is a clever strategy, because the deposit guarantees that the boat companies get some money because some British tourists forget/miss the boats due to hangovers or somehow find themselves somewhere else and are unable to attend. Most times, however, they turn up because they have already paid some of the money and don't want to miss out on something for which they have already paid – even if they are in recovery from the previous night. They then pay the rest of the money and perhaps, if they are 'going to get on it' will be offered drugs from the PR who sold them the ticket. And this is only for the tourists, for running on a weekly basis are special booze cruise parties for the casual workers. In 2012, a Dutch tourist died on a booze cruise from an alcohol and drug overdose and over the summer, two of the cruise ships, when attempting to dock after the sunset at Café Mambo, crashed into each other – both captains were drunk. Despite this, there is no law governing the sea; it is out of bounds. The State has no power. Equally, the potential to intervene is also almost non-existent when it comes to the private hotels/beach clubs.

Private hotels/beach clubs

Over the last few years, there has been an increasing prevalence of private hotels which double as exclusive-looking beach clubs. In this study, the main private hotels are Ibiza Rocks and Ushuaia and the main beach clubs are Ocean Beach Club Ibiza and Blue Marlin.[6] As many of the council workers, health professionals and *Policía Local* said, the introduction of these exclusive locations serve very much to 'continue the party' – in the words of one drug worker, they are *'the new afterparty'*. Let us first examine the private hotels. For several hundred pounds, you can stay in Ibiza Rocks or for an even greater sum of money, Ushuaia. As I will argue, they both cater for different classes of British tourist (see the *Spatial commodification of status stratification*). Essentially, they operate the same sort of concept – enclosed hotel with private pool and music stages where parties/concerts can occur. Security is high, entry is normally only possible through prior booking and, certainly, from the outside, these hotels can seem quite exclusive places to stay. However, the other function they have is to put on concerts and DJ nights. For example, in 2012 Kaiser Chiefs were one of the summer's main coups at Ibiza Rocks while weekly sessions from Swedish House Mafia could be found in Ushuaia. In this respect, it is not only the guests who benefit but people from outside can participate. In this respect, they are a mini-resort within a resort. In Ibiza Rocks:

> As I walk up the steps to Ibiza Rocks, I am immediately attracted to the nearby shops where all manner of 'Ibiza Rocks' paraphernalia is sold. Like the main Superclubs shops dotted around the island, here there is something for everyone – as long as it has 'Ibiza Rocks' plastered all over it; clearly the hotel residents needn't drift too far for either some memento of their break or something to drink by the pool. The well-guarded exterior is manned by a tall, muscular man in an Ibiza Rocks vest. The reception area is grand and appropriately youthful: the floor a plush black and the reception desks where the security staff all have earpieces and pink shirts with 'Ibiza Rocks' on them deal with the numerous young-looking people checking in and out. There are some low, flat plush-looking seats in the middle; it feels like I am reasonably rich and famous to walk in here. I am led by the man down some stairs to the right of the reception and into a foyer which leads out into the main stage and swimming pool area. It is about midday as a scattering of people lie helplessly at the mercy of the sun. The pool is small and one can do little more than get wet in it before getting out. As for the sunloungers, they are back to back

and there is barely a few centimetres between them; they litter the main dance floor area below the stage.

Really this is a resort within a resort. It has everything one could want/need and there is little motivation to drift outside the complex other than to go to the West End or the Superclubs. The hotel offers live concerts twice weekly and promotes the club nights in the reception on a large white board which displays each DJ on each night and the cost through booking at the hotel. It seems awfully similar, yet certainly less classy, than Ushuaia in Platja d'en Bossa. However, by concentrating the groups altogether and even providing them with more reasons to stay in the resort, they amplify their profits; it's to say the blowout is enclosed. [Field notes]

In these hotels, the onus is on retaining the tourist inside and to commit them to spending as much as possible while they are there. This is partly achieved by putting on DJ nights or getting in the world-class music groups but also in the sense that there are bars, restaurants, pool bars, merchandise shops and even mini-supermarkets. Indeed, some of artists who come to these venues have world-class brands which alone are a reason to stay there – and this is why people follow, because they are commercially guided. In fact, there is little need to leave the complex to even score because drug dealers also operate on these premises despite the high security. If anything, the security seems more about ensuring people don't find out too much about what goes on than anything else as confirmed by this drug worker for the local authority:

Drug worker: A couple of years ago, they put restrictions on the after-parties to reduce the number of people partying for more than 24 hours and the number of emergency cases reduced. Then what happened the next year? Private parties started to develop in villas and there was lots of drug-taking so the police intervened a little but what happened last year [in 2011] is that a private hotel opened, an exclusive one and it is open 24 hours [Ushuaia]. It was designed for high-class people and they take what they like in there; there are parties, club nights, gigs, trance parties, etc. They [the managers] have the opinion that it is people's decision to consume drugs and alcohol and anything they do to the limit is their responsibility.

There are also beach clubs such as the Ocean Beach Club Ibiza and Blue Marlin. Normally opening up early in the mornings for people to come and pay for sunbeds, beds or some sort of VIP suite, they also double as a place to continue the party; to continue to drink or take drugs. This, of course, comes at a cost for like the private hotels, Ibiza's market entrepreneurs have done well to offer the same thing to different classes of tourist: the fee is tailored to their spending bracket. The former tends to be popular with a more working-class clientele and the latter being the more upmarket place, and although people go there to do the same thing – get sun, party, drink, take drugs – the separation is therefore only reflected in the price:

> As we walk up the pebble drive to Blue Marlin, the lights start to sparkle as dusk settles in; some skinny young women pose in bikinis by the sofas provided outside. When I go in, I try to go and sit down in the sofa area and am told I can't because it is private. Then I try to make a pass into the dining area but three security men with earpieces descend from nowhere and suddenly surround me: again I am prohibited because it is private. Instead, I do exactly as I am told and sit at the bar; clearly I am not rich or important enough for Blue Marlin. The reception is kitted out with all manner of white, minimalistic furniture under a dim glow of romantic lighting. Generally the clientele look very rich, dressing in shirts, shorts, flip-flops and sunglasses; the day party must be concluding. The cocktail bar at which I sit makes me feel like I am a celebrity; if only I had the money of a celebrity as I look at the prices of the cocktails. I ask for the cheapest, a Cosmopolitan which is only €15, and sip very, very slowly. As I look out into the main area, I see the DJ moving from side to side and two bored-looking, scantily dressed young women dressed in bunny rabbit outfits look around and dance while having a sneaky cigarette. I try to walk into another area out of curiosity and I am again swamped by the staff and told to return to the bar.
>
> When I am finally led to my table far away from the DJ, I pass the numerous private areas which are mostly near his set. As a private yacht gently rocks from left to right with its main light shining brightly in the dark, I look around to see that most of the clientele seem to be Italian with a sprinkling of German and a few British. The staff seem mostly Italian with some Spanish and a few British. Looking through the menu, I see champagne ranges from €25 for a glass to €1,800 for a bottle. Near to me a German man lurches over a waitress at a nearby table of about 12 people who seem to be in their early 30s. When they come to the conclusion of their evening, one

man gets up to pay, pulling out a roll of thousands of euros; he pays the €1,300 bill as if it was pocket money. [Field notes]

During the day, people congregate in these areas to continue the party/start the party – often from an early morning. While their spending is more concentrated and guided in these spaces, so too is their concept of what else there is in the resort area and even the island – indeed, the fact that tourists now spend money on booze cruises or perhaps don't even leave hotel complexes, beach clubs – it's hardly surprising other forms of tourism have been massively crippled and are fast drying up (Chapters 4 and 6). And when tourist spending starts to become unequally distributed, other tourist spaces need to diversify to make attractive the need to spend – and this is when the campaign gets desperate. I want to argue in the next section that the most volatile place where this battle for tourist spending takes place is in and around the West End.

West End

One of the major changes to the nature of tourist consumption practices in San Antonio was the introduction of all-inclusive hotels. As we have seen with the private hotels and beach clubs, the all-inclusive hotel seems to me to have been about restricting tourist movement and thus concentrating, and importantly, containing tourist spending. When this started to happen in San Antonio, at the turn of the twenty-first century, this put pressure on local bars, clubs, cafés and restaurants to look for more coercive means to generate money. This inadvertently occurred in a number of ways but the main three seem to be:

- The proliferation of alcohol deals (such as 2-4-1 but even this has evolved to a variety of offers);
- The reduction in quality of alcohol/increased need to water it down;
- The increasing demand for PRs to persuade people to drink in certain bars.

Beer seems unpopular and the young men say it leaves them feeling (and looking) bloated – which does nothing for their image. Instead mixers and shots are the norm. Red Bull is popular on the West End and often used, as well as cocaine, as a 'pick-me-up' while drugs like ketamine are used towards the end of the night as a 'put-me-down'. These field notes should give you an idea of how these features combine on a night out on the West End. Note how the alcohol deal seems to improve with the more people I bring to the bar:

Dan:	So late in the night [at 4 a.m.]. What deals can I get in here?
Muscled-vest man (MVM):	[Puts his arm around me] Pint of Bacardi and coke, super strong, and one shot of sambuca, tequila, and absinthe all for €7. Can't beat that.
Dan:	Shit, so late but perhaps so worth it.
MVM:	I know.
Dan:	Let's say there's more of me – I know some people down there who might be attracted to that.
MVM:	There's more of ya? Buy one drink each and I will give you a free bottle of champagne if you're celebrating or a free jug of sex on the beach for chilling out.
Dan:	Jesus Christ.
MVM:	That's if there is five of ya and if you are getting wrecked, I will give you a free bottle of peach schnapps.
Dan:	And the one drink is a spirit and mixer – or basically a pint of it.
MVM:	Yeah.
Dan:	And all I have to do to get my free bottle of schnapps is tell you 'I am getting wrecked.'
MVM:	Yep and I will get you annihilated.
Dan:	Even if I brought just women over to attract the guys.
MVM:	Nope, but you can shag them in the toilets if you want and if you're lucky the DJ will film you with his camera!

The offer seems to get almost sordid towards the end as I lose track of precisely how much alcohol I am being offered. But this is because the players in this scenery are interested in my spending and, should I do so, make a trade-off that this will tempt out the darker side of my individual fetishes (Chapter 3); the rationalisation being that 'I could be persuaded if something both extreme and different were to happen to me which I could tell other people I did', for it could be a great story, something to tell the lads back home. Or even if it didn't happen, it could be something good to tell anyway. The offers on the West End are fairly standard as the night begins and have been for the last few years – €10 normally buying a combination of two mixers, two cocktails and a jug of cocktail to share as well as a few shots. However, as these field notes indicate, generally, as the evening proceeds, the more likely better deals on alcohol can be secured as the PRs seek to recruit as many people into the bars as possible. Because money is at stake, it means the PRs look to connect with the punters as soon as possible to direct them into the bar so they will get commission; after all, most are on commission-only contracts – no punters, no wage. This makes them persuasively flirtatious and almost borderline physical when it comes to getting people through the door – regardless of what they are selling or how potentially harmful it could be to the tourist. Yet there is almost no time for them to secure business, as they spend just seconds bouncing between potential groups as they walk up the West End drinking strip.

A similar process occurs for the PRs working outside the strip clubs. They approach only male groups, offering free entry to the strip club and charge around €10 for two mixers and a bottle of champagne to share. Inside the strip clubs (See Figure 8.1), the emphasis is once again made to coerce the tourist to spend. Women are often manhandled into dancing for the scattering of male groups who go in and pretty much sent to persistently seek out each male punter for they too are on commission only, earning money from not only dances but if they can get the tourists to buy alcohol to drink. They may not offer 'extras' in

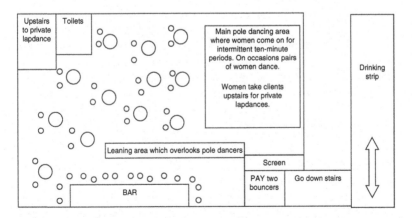

Figure 8.1   Inside a strip club on the West End

principle but if they are desperate and have not earned enough money, it may become a necessity (see Briggs et al., 2011b).

This is all augmented by the lack of regulation on the West End – over the past couple of years, the safety and control of the drinking strip is increasingly left in the hands of the private security guards and the intervention of the police has been minimised to emergency cases that require some official presence. This means the police rarely patrol the space but make intermittent scoping exercises on their bikes in the square below the West End. This leaves only the private security guards/bouncers in charge of problems and, even then, they aren't very understanding when it comes to sorting out disputes, in the main resorting to all-out violence if a situation occurs. Nevertheless, the bar owners can't be too fussy with problematic punters whom, on one night they will ban, but on another, will welcome back with open arms. They can't afford the potential loss in business as they must take advantage of these busy, summer months. Here, in this conversation, one bar owner had thrown out one young man for fighting the night before but let him back in the next night as he was spending money. This seemed to be his default strategy:

Dan: How can you cope with us British?
Ahmed: Some ones are like animals but some ones are like you, like gentlemen. We live with British people here, without British, this island goes down.
Dan: So you need us. You have no choice. But there are Italians, Spanish, Germans here. What's different from the British?
Ahmed: Well, all the Spanish don't like going where the British are, they hate it. Like San Antonio and Bora Bora because it is full of British. But mostly in summer, is mostly British. I love British people [puts his hands on my hands].
Dan: I can see that. So what do you do with people who misbehave in your bar?
Ahmed: I work with them, talk to them, make them happy, sell them drinks, we sell a lot.

This highlights the intense pressure that bars, clubs and hotels in San Antonio put themselves under to secure tourist spending during the summer months; however, as we will learn, this is done at the expense of the tourists themselves (Chapter 9). Sadly, in this process of *hyperconsumption*, and as if it wasn't obvious thus far, young British men have come to believe that women fall within the boundaries of commodities which should be got/purchased; they become 'things', 'objects'. Typically on one night:

> My interest grows in the nearby bar where the music has been turned up and a large number of young men dance around all over the place, half-naked. I start to hear all manner of noises and cheers as I walk into the bar; in front of me stand the large screens and a pack of boys who crowd round one young girl. They chant *'get your tits out for the lads'* as they surround her like a pack of wolves. She laughs a little and exposes what she has, much to their enjoyment and a few dives are made for her breasts before they are tucked away in her bikini top. This she does not enjoy so much and dishes out hand slaps. In general, the half-naked youths continue to jump around on the pool table, scream at each other, high-five and have sweaty hugs – that is until the number of women dwindles and they then retreat to another bar. I stay sitting and a young couple come along, stand in front of me and kiss. His tiny frame can't quite reach around her body as they seem engaged in one of those never-ending teenage kisses which seem to go on forever without breathing. When they peel away from each other, he takes himself off his tiptoes and wipes his mouth. They don't even seem that interested in each other.
> [Field notes]

The function women play – both tourists as well as workers – on the West End tends to be objectified; that is, young men consider young women to be part of something which should be consumed in the environment (Rojek, 2005). After all, they are on holiday and should, most men think (even if they have partners at home) 'shag as many birds as possible'. As we will come to see (Chapter 9), San Antonio as a space displays women as glamorous sideshows to the holiday experience. However, women appear as much in the perceptions of the men who live/holiday there endlessly as they do in the marketing and promotion of Ibiza and its nightlife hot spots. There is almost no escape from this on the beaches during the day, as semi-naked women march up and down with flags promoting the Superclubs and DJs. Women are even persuaded to use their gender to dance in bars to coerce the men in. The West End therefore reflects crude definitions of masculine and feminine ideological gender relations (Andrews, 2005; 2009; see also Skeggs, 2004). In almost every way, women are sexualised; they can be purchased – as strippers, lapdancers, and prostitutes – and it is this transaction which objectifies them and this even happens in the toilets in the bars and clubs:

> As I walk into the toilet, an African man, sings away to himself – he earns money from tips by keeping the toilets clean and offering fresh, manly sprays to attract the 'punani' (West Indian word for female sexual organs or slang meaning pussy):
>
> *African man*: You well, man?
> *Dan*: [while urinating] Yes, thank you.
> *African man*: You want some pussies? I love dem pussies.
> *Dan*: [I continue urinating and start to read the sign above the urinal about punani] Er ... Yes [Sounding confused]. I am here for the punani.
> [This seems to trigger him and he starts to sing]
> *African man*: Go punani, go punani. Punani, punani, go punani. Do do do, de de de. Wash yor finger for de minger [ugly girl]. [And to the tune of the song 'Feeling hot, hot, hot'] Freshen up, up, up for de pu-na-ni. [And to the tune of 'London Bridge is falling down'] Thinking about the pu-na-ni, pu-na-ni, pu-na-ni, thinking about pu-na-ni, pu-na-ni. PU-NA-NI. [Some more young men enter the toilet] Freshen up, up, up for de punani [the other young men laugh]. Wash yor finger for de minger.

While this may be the tipple for the lads up the West End, it is equally the fancy for the rich and wealthy who have been known to hire prostitutes for private parties in villas and on yachts. So, what is for one group, is also for the other. They are the same but how is difference, distance and separation established? As we have started to see in this chapter, this is through the strategic commodification of space which, on one hand, allows the social classes a reassuring detachment, while on the other, creates an intra-class strain.

The spatial commodification of status stratification

In Chapter 6, I discussed a status stratification process which was engaged through conspicuous consumption (Veblen, 1994) and allowed different social classes of these British tourists to ideologically get one over on the other and the masses in the same place. I mean anyone can go to the West End and they could easily be in Malia, Ayia Napa or Faliraki where exist similar spatial concepts – a dedicated area for excessive consumption. So I argued a level of social distinction takes place (Bourdieu, 1984) even though everyone is pretty much doing the same thing they have done on the other holidays (Chapter 6). I said that those returning to Ibiza seek to establish themselves in different areas of the island (Platja d'en Bossa over San Antonio), attend 'better' clubs where their spending power does the talking (Pacha and Amnesia over Es Paradis and Eden) and may even dare to 'live the dream' and work in Ibiza. However, the 'haves' don't want the 'have-nots' to 'have' – even if they may not really have it themselves. This is why these processes of social distinction cannot be seen in isolation; that is, it is not only that people are socially constructing this separation as an individual foreground because elements of the commercial background are also moulding these beliefs. Generally people in my cohort don't seem to realise that they are paying more money to appear higher on this ideological social ladder.

So the other commodification process, which complements that of 'status', is that of tourist spaces. For example, the working-class British tourists in Ibiza Rocks think they have one over on the regular working-class tourist because they are staying in 'Ibiza Rocks': the 'good life' is on their doorstep, it costs more and the brand name alone has with it attached a degree of social kudos. Yet they are part of the same class bracket. The generally upper working or middle class residents of Ushuaia in Platja d'en Bossa think they have one over anyone in San Antonio because they have more money to spend, Ushuaia has only the 'best' in house parties, is more expensive and 'classier' than any hotel in San Antonio. But everyone in these places are all doing the same thing – engaging in *hyperconsumption* so the common denominator in all of these examples is expendable capital. With money, one can rise up the ideological social ladder – albeit temporarily – these tourists can feel special, feel like one of the elite, spending thousands like Mario Balotelli in Ocean Beach Club Ibiza or throwing money at the party lifestyle like Tulisa Contostavlos in Ibiza Rocks (Chapter 5). Better 'Ibiza experiences' come at a *price* but also in a *space*, and in paying that price and by being in that space, one can distinguish themselves from the 'other' – these tourists can achieve elevation above the masses (Bourdieu, 1984).

However, this importance is expertly exploited by the global corporations, market entrepreneurs and tourist organisations and this as evident on the beaches as it is in the Superclubs and beach clubs. For example, the tourists lying on a beach towel on Bora Bora beach tend to look a little inferior to those who have paid money to get a sunbed. However, those with a sunbed are put in the social shadows somewhat by those with a sunbed and umbrella. Yet above them all are those lying all over each other on VIP four-poster beds; they are on top of the world. In this way, 'VIP' is stratified (VIP bed) which makes the most basic paid selection (sunlounger) look a little socially inadequate. Who would then go to Bora Bora to lie on a towel on the beach? Gonna look a bit stupid, hey. Better get a sunlounger at least. But we are on holiday, why not get a VIP poster-bed? After all, this is 'Bora Bora' and this time won't come around again (Chapter 6).

In all the Superclubs, there are VIP areas where the 'special people' can hang out because they have the capital; they don't really dance but sort of hang around looking at people to see if they are looking back at them in envy. This *spatial commodification of status stratification* takes on another meaning in some of the beach clubs where VIP has various levels as we saw earlier with Ocean Beach Club Ibiza. Take Blue Marlin, for example (see Figure 8.2). As I noted earlier, one cannot just walk into this place without a reservation. On walking in, one immediately sees there are two main VIP areas, ranging from those behind and in front of the DJs and dancers, conveniently next to the glamour of the champagne bar. This is the most exclusive part of Blue Marlin. Next, the VIP spots to the right of the DJ, near the private dining area denote another zone of exclusivity which makes standing at the bar or on the dance floor with nowhere to go look pretty inferior. Below, things descend into a kind of poverty on the rich scale as the 'regular diners' (like me) sit on the beach area. During the day, the prime spot to be is in the 'VIP lounge bed' but if you haven't got the cash, a VIP bed may do – even if there isn't much space between them. But if people are obviously here for the first time and didn't anticipate the cost – perhaps taking the *short cut* (Chapter 6) – they'll get a sunbed barely centimetres from the next, and it'll be obvious they are a novice because they have no money to show for it.

Thus status stratification is commodified and spatially determined. Different classes of Brit can feel they have got one over on the 'other' which occupies the same territory by spending that little bit more and looking that little bit more special. It is therefore necessary that the NTE space reinvents itself to establish a new elite where the 'have-nots' can't go. Someone doing that is way out of their depth, much like the young working-class Londoner we met in Chapter 6 who went to Ushuaia; without spending power, he looked massively out of place yet vowed to return when he had the capital. The same goes for the Ocean Beach Club earlier in the chapter whereby different levels of VIP separate the

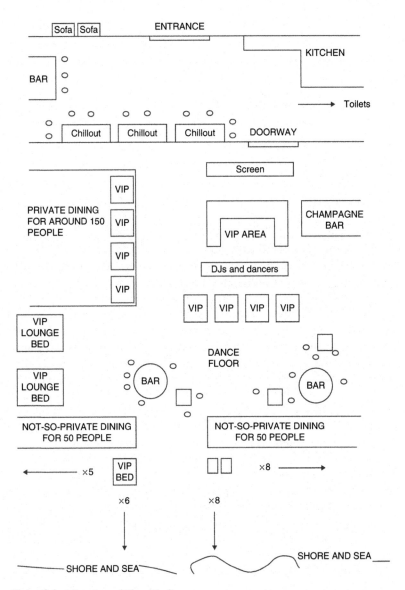

Figure 8.2 Mapping of Blue Marlin

'players' from the 'wannabees'. This is one reason why people want to come back to Ibiza – to do it properly, to have the money so they needn't worry about what they spend, just as long as they retrieve this status. This is the commercial success story which is going on, but, as we will see, spending drifts into the realms of the ridiculous and often to the detriment of the tourist: the winners – the global corporations, commercial entrepreneurs and marketing experts – profit at the expense of the losers, the British tourists (Chapter 9).

Conclusion

This chapter shows how a number of important key structural areas specific to Ibiza play a part in the deviant and risk behaviours of British tourists. Firstly, it has highlighted how business interests, and even elements of the island's infrastructure, are now almost incestuously bound to the tourism the British provide; in many ways, they operate in symbiosis which is so embedded, that it becomes difficult to see an alternative system. On the other hand, it shows the same features of the local economy are totally under-resourced, lack strategic governance and general investment, and don't seem to benefit much from tourist spending – this is called 'leakage' (Opperman and Chon, 1997); when tourist spending does not seem to directly benefit the economy of the destination. This was certainly the case for the overburdened health service and the police. Generally these areas of the economy are poorly equipped to respond to the type of tourism which is incumbent in San Antonio and certainly find it almost impossible to challenge the complexities of how the informal economy has advanced (prostitution, sex industry, booze cruises). And even when there is some effort to reduce the problems, the corporations, commercial entrepreneurs and the Superclubs just walk all over the new laws, finding other ways to generate profit. My data also shows how the concentration of spending in spaces designed for excess reduces the tourist's decision-making power and instead encourages *hyperconsumption* and deviant and risk behaviours. However, this appears to the British tourist as new 'experiences', and they are sold at a premium so that people can feel more special than the 'other' – even though they are doing the same thing as everyone else. It seems that everyone is surfing the commercial wave in Ibiza but the real wipeout takes place against the local businesses and the tourists themselves. The next chapter shows exactly how this takes place, with a focus on remaining part of the Southside Crew's holiday.

9
Capitalismo Extremo: Risk-Taking and Deviance in Context

After no fewer than six different PRs approach us with offers for booze cruises, bars and clubs, we stumble out of the bar at around 4.45 p.m. and they permit me to continue recording. We walk slowly down the road, distracted by different attractive women trying to pry us into bars or sell us tickets to help shape our 'night out'. As the sun bounces off our half-naked bodies, Jay practises his kickboxing techniques on me and shows me different methods of killing people. Nathan then whispers in my ear about how he 'pulls women', telling one PR woman he will 'fuck her senseless in his apartment.' '*Charming*' she replies. Their blunt advances continue with little success:

Jay: Nice arse.
[The girls passing by walk on unimpressed]
Nathan: [To me] Mate, you have to push the birds out here.
[One PR woman approaches us]
PR woman: Hi guys, we just want to let you know what is going on Thursday ... [Hands out leaflets for a pre-party booze cruise]
Nathan: Nice face. Pretty.
Jay: Then to our hotel to party or you could come pre-party?
Nathan: We will fucking smash you up, fuck you, all sorts.
PR woman: Eugh, that's disgusting. That's vile.
[All the lads laugh while Nathan remains serious about the offer]

Introduction

Here continues the Southside Crew's first night on holiday in San Antonio. All the data presented thus far has been from their lengthy focus group discussion. They are now loose in the social context and ready to continue the party. In doing so, and as this short excerpt shows, their no-nonsense holiday intentions and fetishes surface in tandem with the relentless bombardment of offers to help shape their night out and their holiday in general – all of which involve spending money, drinking, drugs and are sold with the ideology and promise of sex (Andrews, 2005). In this chapter, we will see what happens on their first night and how their attitudes towards a holiday blowout are persistently coerced by the commercial and commodified elements of the social context (Chapter 8), which results in deviant and risk activities.

Therefore, the idea of this chapter is to put in context everything which has been presented thus far by using case studies to exemplify moments in which deviance and risk evolve in the San Antonio area, as well as in other tourist hot spots in Ibiza. I think what we are seeing in resorts like San Antonio, and across other tourist spaces in Ibiza, is a form of extreme capitalism or *capitalismo extremo* – a sublime money-making process led by global corporations, commercial entrepreneurs and tourist companies/organisations which ideologically pied-pipers these tourists to 'seize the moment', 'live the dream' and engage in excess, deviance and risk – all at the expense of themselves. And while it's fun to come on holiday, have a drink or two and spend money, *capitalismo extremo*, I want to show, takes no prisoners because it is concerned only with exaggerating these blowout attitudes to make profit – even if it means significant personal, financial and/or social loss for the tourists. Unknowingly, the British tourists in my sample participate in this social system, which often leaves them penniless and to their detriment, yet this doesn't seem to matter much to them. This is because their *habitus* and ontology are already preset on reproducing similar kinds of behaviours back home (Bourdieu, 1984, Chapter 5), so to do them abroad, when they occur in an exaggerated and acute manner, means they are even more enjoyable, even if they may appear from the outside to be bizarre and extreme. There is even pleasure in the painful consequences (Lacan, 2006).

Space, deviance and risk: The Southside Crew on the drinking strip

Let us first start with how holiday attitudes towards a blowout and the commodified space combine to reproduce *hyperconsumption*, and, as a consequence, deviance and risk behaviours. In what follows on their first night in Ibiza, like many in the sample, the Southside Crew are keen to go hard at it. Note how there is a persistent battle for their business (Chapter 8) as they are incessantly stimulated by the offers for sex, drugs, drinking, clubbing, and generally anything. The main thing to point out at this stage is the way in which the ideological discourses of 'sex', 'booze' and 'best night' are reinforced by the PRs and, it is my contention, that this thrusts them into thinking that they are precisely the things they should be getting from the night – one way or another – and the only way to achieve this is to spend money. Just like the stag parties (Thurnell-Read, 2011), the way in which masculine identities are played out here defy the hard and controlled male body, and in the time and space which is the holiday and resort, a less controlling masculinity is enacted; one which is as extreme as it is self-mocking, and must be sought through excess and overindulgence to mark the occasion (Featherstone, 1991). Their treatment of women in the same arena is also objectified (Chapter 8).

The Southside Crew are here!
It is about 5 p.m. We walk down the Bay area, half-naked with T-shirts in hand as if we are kings of a new land. The Southside Crew discuss how 'black birds try to offer them blow jobs' in Magaluf and recount other memories which involved deviance and risk:

Jay: We get to the hotel and he [Nathan] is minging, drunk five pints if that, he is smashed and he gets up there on the first floor, strips off naked, gets on the edge of the balcony and says 'I'm gonna fucking jump' and everyone is down there saying 'jump, jump'. It is 4 a.m. in the morning and we just got here. We were only in the hotel two hours and we had been kicked out with a verbal warning. He ended up out by himself, three of the five nights. He was fucking on so many pills.

They then discuss cheating on their girlfriends and partners and agree that 'what she don't know, don't hurt' and the conversation takes a reflective turn when Jay recalls how he spent in the region of £130,000 when he was dealing and using cocaine. There then follow accounts of prison time, yet despite these reflections, we don't seem to have moved far since leaving the bar and the same music still pumps out next to us. We are stopped by another PR man working for a club and a couple of them listen in on the deal. Thirty seconds later another PR couple approach us with tickets for Eden and Es Paradis as one starts to wrap around us free-entry wristbands as if to indicate we may end up there – and if we do we can go in free of charge and spend. Apparently Mark Wright (from *TOWIE*) is one of the main reasons why we should go (Chapter 5) … as well as the potential for *'pussy'*:

Irish PR man (IPR):	Plenty of pussy in Eden tonight.
Jay:	Fuck it, I'm gonna pay for everyone.
PR woman (PRW):	I'll sort your tickets out then because we also have Professor Green. It's gonna be really good guys. [IPR imitates shagging a girl from behind and laughs about how he claims he pulled two young women in a night]
PRW:	I guarantee you, you will pull in there tonight.
IPR:	Oh mate, you gonna find some hot pussy in there [laughs to himself]. Want to see my cock?
Jay:	I'm gonna go, I'm paying. I don't give a fuck, how much. Tell me. I have unlimited money.
PRW:	So you get the bar crawl and entry into both clubs. The water party comes on about 4 a.m. in the morning and that's when the fun really starts.
Jay:	Let's PARTY!
Dan:	You can't pay for me.
Jay:	[Hugs me] Come out please, let's go. We have to!! I think you're a top man. I don't know why I like you mate, but I do.

We eventually drift towards the town centre. The boys debate going to a club now but there is general confusion about what should happen next. We are approached by some more PR women and asked where we are staying. They try to sell tickets for the Zoo Project which operates on Saturday but this doesn't stop Paulie from having a flirtatious exchange with them.

Time to eat … properly

We eventually settle in a British-style café in the West End. The waiter comes over and gets upset because we have our T-shirts off so we put them on. With little thought to his mates, Jay orders everyone a shot of tequila each, a shot of sambuca, a pint of beer and a litre and a half of sangria between us. The waiter stands there unimpressed by our brash behaviour, and as the attractive women walk past, they are stimulated into wolf whistles and yelps of *'nice arse'*. When the food comes, Jay seems a little too drunk to hold his cutlery so Marky cuts up the chicken and feeds it to him. Jay then pours more drinks saying *'let's smash these'*. As we finish more shots and drink, yet more sangria is poured for us to sink. The sambuca is next and we knock them back with surprising ease. They agree to go to the hotel, change, drink and then hit the West End. Jay hugs me, saying 'I like you, mate. Come back, listen to music, chill, have a few drinks'. Although we have finished our food, there still remains a substantial amount of alcohol on the table. The shots have gone and most of the sangria but the pints remain intact. Perhaps feeling guilty that we have neglected them, they down as much beer as possible and slam down the pint glasses. We leave the restaurant and again, the order of the night seems to be changing as they are stopped by more PRs telling them how much 'pussy' is in their bar/club on the West End.

To the hotel

As we get closer to the hotel at around 8 p.m., Jay decides to do something extraordinary. He takes off his shorts and runs naked as fast as he can towards the hotel. When I get to Jay's room, the door is slightly ajar. Knocking, I enter, and see him lying in a pile on the bed while a small, paint-stained CD player lamely hammers out music. I wake him and he slowly comes to, only just recognising me. I climb over the beds to the balcony. He tries to sober up and get himself together for the night out and in doing so, stumbles over to the safe to get the rest of his money but has forgotten the code. Frustrated he reaches to a zip pocket on his shorts where he hopes to find some reserve euros. Some time later, his friends then enter; Paulie is ready to go and invites me to his room which he shares with Marky. He shows me his CDs and puts on a little tune and does a little funky dance, while he checks his hair in the mirror – they both seem quite sober despite the volume of alcohol they have drunk. We all then walk to Jay's room but when we enter there is a smash; he has dropped a half-full glass of Disaronno. He reasons *'fuck it'* and the others, seemingly tired of this, say they'll meet him downstairs. As we leave, Nathan then comes in after a shower; he is dressed in all white shorts and vest.

While we wait for Jay to sober up and get ready, Paulie gets impatient and I go down with him to the all-inclusive bar, which is littered with young people, sipping from plastic cups and ready to go out. The African men continue to try to tempt people with sunglasses and other meaningless holiday gadgets. Paulie goes off to look for Marky and Jay so I remain with Nathan who tells me the approach to drugs here is too soft and the police need to arrest the *'black people who are the dealers'*. We have a few beers while waiting for the others before Jay then stumbles into the conversation and outlines the plan for the evening. Even as we sit down, PRs continue to approach us advertising Es Paradis and Eden even though we are displaying free wristbands for club entry. Again the conversation revolves around the potential for *'pussy'*:

PR guy (PRG1):	€40 gets you into two clubs and there will be all the women you could dream of.
Nathan:	I am getting hard already.
PRG:	Best thing is to buy these tickets, you lads, get down there, get loads of pussy.
Dan:	[sounding suspicious of the promises] Loads of pussy?
PRG:	Fucking loads of pussy, mate.
PR girl (PRG2):	There really are so many girls in there.
PRG:	I swear, 80 per cent of what I have sold have been to girls. I swear, loads of girls.
Paulie:	We'll see what we do.
PRG:	It will take you three hours in the queue, I know full well.
PRG2:	You will have to queue.
PRG:	What I could do is throw a water party in for you as well.
Dan:	But there is a water party on anyway.
PRG:	Yeah, in Es Paradis but we can throw that in there for the €40.
Paulie:	But we have free wristbands for Es Paradis.
Dan:	We have the free water party.
Paulie:	We'll have to see.
PRG:	Up to you, but we'll hopefully see you in there, hey. [They leave]

To the West End

We argue over whether to walk or take a taxi only to be approached by another PR man who is offering a Professor Green pre-party at Eden for €20. Marky returns with six beers and we are forced to down them quickly because the others have decided we are leaving. As we walk on the road, attractive women continue to parade up and down and Jay shouts 'OH MY FUCKING DAYS, SHE HAS GOT A TIGHT ARSE! *Jesus Christ.*' He then concedes to having €230 to spend for the night before reminiscing that the last time he stayed here, when he was smashed drunk, he *'head-butted a girl'* and then laughs. We leave for the drinking strip. On the way, Jay strays into the road without looking and I have to shout at him to get him back onto the kerb. As we pass the bars, the music booms out across the small road. The night has begun. On the way down, we are separated and tempted by new offers for drinks and entertainment. But it seems Jay is hungry and stops off at KFC. Even as we come out, the offers for drink, deals, drugs do not stop for one minute:

Jay:	Doesn't matter as long as you are minging. It's going to be a fucking messy night.
	[Jay stops to get KFC – his third chicken dish of the day.]
Jay:	In the room I was fucked-up.
Nathan:	I am not feeling good now, I need to sober up. I'm not drinking until I get West End, mate.
Jay:	Did we pay in that other place?
Paulie:	YES!
Jay:	Man, these people [African men] live off tourists. I have €650 for three nights.
Paulie:	I have €200 per night.
	[An African man approaches me]
African man:	Coke? Hash?
Dan:	No thanks.
	[Seconds later]
PR man:	What club are you doing?
Jay:	SOUL CITY!
Paulie:	Don't know yet. We are meeting 14 mates.
	[We walk on and as we pass Lineker's and another PR man approaches]
PR man:	We are doing Clubland, fucking Professor Green, all the DJs. Come in here, get messy first. €10 for two double spirits with a mixer, shots of sambuca and a bottle of tequila if all four of you come in.
Paulie:	Want to go in?
	[The group seem torn and no one looks motivated to make a decision]
PR man:	Come in boys, it's not shit, it's fucking superb. Maybe able to get you champagne if you want. You know the birds in here. If you don't come in, try us tonight. Lots of fucking British birds.
	[We move on and again pass some African men offering drugs]
African man:	Do you want to get high, mate?

Dan:	No thanks.
	[Music booms out]
PR man:	Guys come in, two shots ... [but we walk out of earshot because there is no one in the bar]
	[20 seconds later]
PR man:	Guys, €5 for two pints of San Miguel.
Paulie:	Nah.
	[50 seconds later]
PR woman:	€40 for Eden otherwise €45 on the door. Coming in guys?
Paulie:	Is there any gash?
Dan:	What's 'gash'?
	[They all look at me unforgivingly]
Marky:	It's fanny, Danny!

The West End

When we arrive on the West End, we reluctantly enter Soul City at about 9.30 p.m.; it is empty. We were promised all manner of women by the PR who offered us free shots but there is no one there. Jay is disappointed by the empty dance floor and it almost kills the atmosphere. The music is too loud and I can't establish the deal on the drinks and instead a beer is thrust into my hand. We can't make easy conversation without pretty much shouting directly into each other's ears. We raise cheers and sink our shots and another is thrust into my hand.

Unhappy with the lack of popularity, they move outside where they are told they cannot drink there so drift onwards up the drinking strip. Outside we meet the IPR from earlier in the evening. He slaps their backs and tells stories about '*shagging women.*' Once again, we are offered tickets for 'Clubland': a deal with access to Es Paradis and Eden. We manage to prize ourselves away from him but he catches up with us and directs us towards another man further up the strip. We are then offered laughing gas for €5 by two women wearing next to nothing. Nathan wants to go back down the strip but Jay moves in for some offer for Clubland and, when I can't see Paulie, it feels like everyone is going their separate ways.

Halfway up the strip, Marky is tempted into a deal for €40 for a bar crawl led by beautiful women – on the bar crawl they must pay for their own drinks but get free access to both Eden and Es Paradis (where they are told there will be a water party, loads of 'pussy' and a wet T-shirt party). We already have free entry to Es Paradis because of the wristbands so we are really paying €40 for semi-naked women to march us around, show us to a bar where we buy our own drinks and then let us into Eden. Marky and Jay seem very keen and move to muster up support from the others. When Jay can't see Paulie or Nathan, he offers to buy tickets for them. The man he deals with seems completely wired and his eyes move around like jolty discs; he takes some money from Marky and more from Jay, then hurriedly puts €120 in his pocket, taking little care about folding the notes. When Paulie catches up with them and realises they have bought tickets for a bar crawl in which they have to pay for their drinks, he is not happy: 'oh for fuck's sake man, a walk around a few bars' he says sarcastically. There are some awkward moments as Marky and Jay try to persuade Paulie that it is worthwhile because they'll be led around by pretty women. The PR man talks it up and immediately tries to usher them into where the bar crawl starts. Jay tries to convince me how good it is and the PR man talks it up: 'It will be the best night of your life, mate!' Paulie lingers around looking frustrated. After a few minutes, I go into the bar where I am welcomed: as they hug and high-five me, we all get a drink and I end up dancing in puddles of alcohol and sick on the floor.

I continue to jump around in the puddles and celebrate with my new friends. The DJ can't help but mention Mark Wright, the character from *TOWIE* (again!). I am in the pre-party for Eden and Es Paradis for free. There are some resident dancers erotically dressed in red, grinding around on the speakers; they look extremely thin and unwell. The lads look at them and there are some flirtatious exchanges between them. I continue to jump around in the puddles and my flip-flops fill with the mixture of vomit and alcohol. Jay sticks his tongue out at the resident dancers suggestively. The DJ periodically shouts:

DJ: MAKE SOME FUCKING NOISE!!!!

There are some high-fives and hugs which include me. Marky says 'it's gonna be a good night' but then realises the shit he is dancing in; 'look at all this fucking dirt' he says and lifts up his new shoes in disgust. We are led out of the bar by the women in bikinis and high heels. They all hold flags as if to symbolically lead the way for the debauchery which is to follow. The lads follow mesmerised as if the Pied Piper is leading them with a melodic tune. As we move to another bar, we lose Jay but I find him quite happily and aimlessly wandering the streets. As we enter the second place, we struggle to get to the bar. There is some wait and meantime Jay hugs me:

DJ: LET'S GET FUCKING TWATTED [drunk]!!!

The girls who have been leading us around thus far start to dance with the lads, who reciprocate by grinding their bodies up against their skinny, bikini figures. Paulie is especially keen on this. They rub up and down against the lads until it is time to move on to the next bar. It is now about 11 p.m. The same female dancers gyrate on the tables around another bar and seem to completely mesmerise Nathan. There is barely time to have a drink before Professor Green says that we should move on. The red-dressed female dancers go to the street in formation with a pair of big banners and a security guard. We follow them and reach the next bar, where some young women walk around selling shots and others have a kind of booze-sprayer which they try to foist on people. While the others dance, Nathan chats with one of the booze-sellers, even getting to hug her and touch her bum a little. But when she sees that he is not interested in buying anything she quickly moves away. On the dance floor, people crash together, jump around and everything is rather chaotic.

As we are led away to the next bar by the red-dressed females, the group breaks up along the drinking strip. Marky and Paulie continue in pursuit of the red-dressed females while Jay and Nathan walk into KFC. They quietly eat from their bags of food, which is not a pretty sight; gnawed chicken legs fall to the ground and their faces and fingers glisten with the fat. After eating, they sit on some benches; Jay is so drunk that he almost falls off several times. Then three Brits in their late twenties sit nearby and scowl; one of them says something insulting in our direction and walks over in a threatening manner. There are some angry exchanges but Jay gets up and walks towards the clubs. 'Take a piss' Jay says and stumbles across the street, taking several minutes to unzip his trousers before staggering over to a rubbish bin and urinating in it.

Meanwhile, Nathan continues and is now 100 metres ahead of us. Jay collapses again, this time on a bench. It is only 500 metres to Es Paradis but it takes us about 40 minutes to arrive. They then stand staggering in front of the entrance of the club. One of the doormen gives a sceptical look and shakes his head, but they are permitted to enter. Ten minutes later, the two of them waddle out and head towards Eden, which is situated on the other side of the street. They explain that Es Paradis is 'crap and empty'. We stand in front of the entrance and again the doormen look sceptically at my two companions. We walk inside and there are only around 50–60 guests in the giant club. They each pay €12 for a vodka Red Bull but as Nathan takes a sip of the drink, he tips it over. Then a female photographer comes around, perhaps working for the club. Nathan tries to reach out for her bottom and gives a grunt but she jumps away. Suddenly Jay gets up to leave, having barely sipped his drink. When we get outside we can't see Jay, and Nathan seems perplexed without his friends. I take a seat to take stock, and Nathan disappears. I walk back to the drinking strip but can't find him or any of the others. It is only about 1 a.m.

The next day ...

I find them the next day, late in the afternoon lying on the beach, playing with the sand. They reiterate how their first night was a *'great night'*. So what happened after I lost them? It seems Jay couldn't remember where he went or how he got back to the hotel. He thinks he got in at 2 a.m. but Marky says he *'got lost'*. Paulie stayed with Marky but found some of his 14 mates and got in about 6.30 a.m. Marky left for the hotel but was approached by an African prostitute, had no money, *'ran back to the hotel'* to get some money but then couldn't find her. He finished the night slamming tequila shots on his own until about 4 a.m. He too couldn't remember how he got back to the hotel. Nathan? Well ...

Nathan: Got sucked off by two black birds [African prostitutes].
Dan: Oh yeah? What, down the West End?
Nathan: Yeah.
Marky: Yeah, I almost paid for a hooker on the way home.
Nathan: Well it was going to be for €40 because it was two of them but I didn't have money.
Dan: You didn't have money? So what did you give them?
Marky: Give them dick!!!
Nathan: They sucked me off a bit, I shit myself then I just told them to piss off afterwards.
Dan: [Disbelieving] You shit yourself?
Marky: [Laughing] Too much drink, mate.
Nathan: I don't know how it happened mate, but as I was cumming [ejaculating] I shit myself [starts to smile]. I had shit everywhere, all on my white shorts.
Dan: I wouldn't put that on Facebook.
Nathan: Stuff like that stays on holiday, mate.

Different holiday agendas start to put strain on relationships as the night goes on and the commercial concentration of so many unlimited pleasures (Chapter 8) begin to resound with the individual's need for *jouissance* (Chapter 3); so much so that by the end of the night, the group has completely dissipated. Deviance and risk arise in this process of *hyperconsumption*: this is the first night of the holiday and the liberation begins – or so they think (Chapter 7). As the field notes show, sex was being sold as a commodity in San Antonio – as a promise for capital exchange if one attends the clubs or goes into the bars. Leading this market assault were the PRs who made continuous sales references to the lure of 'pussy' which validates the male consumer view that women exist to be 'shagged': this logic lives in their point of sale, their branding, and the way they commodify it, thereby transforming women from a subject into object (Chapter 8). Therefore while San Antonio seems to represent a gendered landscape (Andrews, 2005) designed for male gaze (Pritchard and Morgan, 2000) where both young British men and women play out their gender roles (Andrews, 2009), payments are made with the expectancy that sex will be the end result. The direct and blunt sexual advances of the Southside Crew often resulted in instant rejection from other female tourists and this appears to redirect their thirst for sexual conquest towards those working in the sex entertainment sphere. If it is not the bar crawls led by half-naked women, there are open markets for strippers, lapdancers and prostitutes. Their concept of masculinity revolves around notions that are both stereotypically hard (Jay showing me how to kill someone) as they are perverse and self-mocking (Nathan shitting himself and getting a blow job) (see Thurnell-Read, 2011).

Capitalismo extremo here blinkers the Southside Crew into thinking that everything they could ever want is in the resort and that 'dreams becoming reality' are just around every corner – if only they part with more money to discover them. The truth is, despite their unrealistic expectations for the night, they have to believe they will have a good time; that it was all worth it. And as the field notes show, even when these kinds of nights go wrong, they are still 'great' – still something to remember as we will see in the next example, for a group of inexperienced girls who took the short cut to Ibiza (Chapter 6).

Learning what's what and what's not in the new adult adventure playground

With perhaps some awareness about what these holidays are all about (Chapter 5) but without having any experience (Chapter 6), I am arguing that many 'first timers' in San Antonio are likely to be the most vulnerable. Take Lizzy and her friends from Manchester who were looking for a *'cheap holiday'* over the summer of 2012. All in their late teens and about to start university in different places, this was to mark their first holiday away together. They had wanted to go on a Club 18-30 but one of their party wasn't old enough. After being advised by a local travel agent that Ibiza may offer them a good deal, they booked seven days all-inclusive in a large hotel near el huevo. They didn't know much about the DJs or some of the Superclubs like Amnesia which *'were in the middle of nowhere'*. While two of the four drank on the plane, when they got there they *'didn't understand what was going on'*. After dumping their bags, they went out but didn't make it as far as the West End, instead partying in a bar no further than 100m from the hotel. After one bought one drink for €7.50, a PR told them, it was best to go with the deals because *'it's not worth drinking sensibly'*. They were a little confused at the onslaught from the PRs but were persuaded by one to pay €10 for tokens which they could exchange in a bar for three shots and two mixers (with treble shots).

Lizzy and her friends hadn't tried drugs (bar one) but significantly amplified their drinking. Lizzy says that while they 'did all the big clubs like Eden and Es Paradis' and went on a boat party with hundreds of people, most nights were spent on the West End getting drunk and dancing. In one bar, while hammered, one of Lizzy's friends jumped off the bar and broke her wrist but didn't realise until she got back to the UK. Whereas at home they tended to go out and return home together, when they were abroad one in the party disappeared a few times to have sex with a couple of men on a few nights; she unfortunately *'caught something'* and, after persuasion from her friends, had to seek treatment from the doctors on returning.

Firstly, look at how the holiday company recommends Ibiza. Secondly, they have no clue about what Ibiza supposedly stands for but are only interested in partying. In fact, this group of young girls are learning first-hand from the PRs what is expected of them on these holidays as they experienced when they went on the boat party, where they saw other people doing the same thing as they did (Chapter 7). They think the 'big clubs' are Es Paradis and Eden but will likely try and make some attempts at the other clubs as they vow to return (Chapter 8). Importantly, however, in the supposed 'liberation' from home life, one unknowingly broke her wrist and another caught something – the subjective desires to indulge come to surface as one disappears to have sex while the quest to retrieve some individuality arises in the moment for another when she jumps off a bar. *Capitalismo extremo* makes it possible for people like Lizzy and her friends to believe that this is the fun they should seek in their youth (Chapter 7), even if it means they suffer some painful consequences (Lacan, 2006).

The consequences of playful deviance and risk: Surreal Brummies constructing surreal stories

The holiday aim for many of the Brits in this sample is as much to have a blowout as it is to create/construct memories. With almost everything riding on one week/two weeks in the year or even a short period of time away from home, 'fun' needs to be had – it is a determined fun (Chapter 5). We have seen how *hyperconsumption* results in haphazard spending and this is where things can often go wrong – especially in a social context where all is not what it seems. Agency can become blurred and many often end up at the mercy of the informal economy and in the hands of the criminal justice system and/or health services. Nevertheless, it's all the same – they were 'great times', and like Lizzy jumping off the bar and Jay running naked towards the hotel, there is some attempt to transgress into the realms of the bizarre and extreme to claim back individuality and 'freedom' (Chapter 7). This seemed to be the case for the Surreal Brummies, a group of young men aged between 18 and 19, who had a number of stories to tell from their first holiday in San Antonio.

For most of the Surreal Brummies, this is their only holiday of the year. Of the 15, two are unemployed and borrowed money from friends to come away, but most are about to start at university (Chapter 5). They have no particular holiday agenda other than to *'have a laugh'*. Only a few have come to Ibiza before, and some of the group went to Malia, on the Greek island of Crete, last year on holiday (Chapter 6). They chose Ibiza because some friends recommended it. At home, they are not really into drugs and confine drinking to weekends. Most say if they go out, it is on Saturdays and a couple say they have only *'one big night a month'*. The few that have girlfriends said their partners protested at their trip to Ibiza; a couple even dumped their partners before coming out because they *'didn't want the hassle'*. One didn't tell his girlfriend he was coming away and left his phone at home. Members of the group have taken between £700 and £1,500 for a week's spending but some lent others money over the course of the trip. I meet them in the hotel with their bags packed for home. As they finish on the video game, they give me some clues to their holiday actions which involved 'hospital business, a prostitute and drinks and drugs' says one. Another says 'If you just come over, you've got to take it [seize it]' and that the fact that 'nobody knows you' means it doesn't bother them (Chapter 7).

They reason they drink every day on holiday because *'at home they can't'* and that alcohol helps to 'get the most from the experience'.

Because they are a large party, they have different interests and different agendas. Consequently, the Surreal Brummies tended to just split up most nights. Different combinations of group members and vague plans to *'see what happens'* on their 'nights out' made for different 'experiences' each night. The more 'experiences' they can construct, the more they can share in the group context. Like most who stay in San Antonio, the Surreal Brummies don't have enough capital to do the clubs 'properly', have no interest in the rest of the island, as they say, so they are confined to supermarket beers and nights on the West End (Chapters 6 and 7). It doesn't mean they don't go but that they can't go as often and certainly can't buy many drinks. They can, however, take advantage of different drinking events – such as the 'booze cruise'. When at a pool party, they were approached with the offer of 'booze cruise' tickets of which they gladly accepted because they *'was on holiday, innit.'* The one they went on cost €60 and included a free ticket to Space. The evening began on the boat at 6 p.m. and concluded at 6 a.m. in Space, but they estimated they spent about €80 per person that night on alcohol. They said it was *'binge drinking on another level'*. In fact the level of intoxication that night played a major part in one of the group having sex with a prostitute (Chapter 8).

While the Surreal Brummies say they didn't have 'women' on the agenda, they said they were generally open to a holiday fling. Perhaps they say this because they didn't have much luck. However, only one in the group seemed to have some luck but he had a partner in the UK. As we have discussed, many considered their relationships and even marriages redundant when they were on holiday with their mates. Bob was the only successful one with women on holiday and this created some envy among the group members. However, shortly after this, the conversation takes a different direction when inter-group relations are bizarrely explored. It quickly emerges that two group members, Stevie and Bob, kissed each other several times in one bar on the West End – something which they said they certainly wouldn't do at home because of their local reputation (Chapter 7):

Dan: When you say kiss you mean full on?
Will: Him there [to Stevie]?
Stevie: Yeah, I got off with him [Bob].
Dan: [Feeling it is some big joke] Hang on, we're talking a little peck.
Bob: No, full on.
Will: [To Bob] You fucking gay!
Stevie: I did it four times because people kept missing it.

Will: When you did the fourth time, we were a little bit worried.
Stevie: And when I got in that night, all I could smell was the aftershave on me.
[Laughing, and Bob and Stevie high-five each other]

This was problematic for some members of the group; a few like Will frown on this behaviour. It doesn't seem to reinforce the ideology of a heterosexual moment which for most in the group is the norm. However, this is the holiday and in absence of regular norms, other excuses to explore relations and reactions to those relations are transgressed. On the contrary, what may appear from the outside to be 'bizarre' feats were also some measure of group social standing (Chapter 3). The former is explored in this next excerpt where masculine feats of endurance and stamina become the order. Here, Bob recounts how, after 'five bottles of Jägermeister that day' he swam across the marina at night ... which is also the primary shipping lane for the port:

Bob: I went in the water bar with a few other mates and when we've thought it's gonna be a good idea to swim across them water bits [marina] and I've gone through them, past the boats and I was like halfway ...
Dan: You seriously swam?
Bob: [Getting excited] Yeah, and I've seen like a speed-boat like the water sports and there was just a banana boat in the back and this big thing and it was all inflatable. And there was this ring thing which you can fit three people in and I thought 'Yeah, I've got to get that away' so I turned it upside down, and me and my mate swam in it and it took about three hours to get to the other side of the marina. Then I brought it in the room and the boat was like it filled the floor, it was the boat and the bed.

Under *capitalismo extremo*, *hyperconsumption* and a drive to transgress *unfreedom* make this possible (Chapter 7) and, as a result, stories of amazing feats become legendary among the group and Bob receives praise for his heroics. Even when he brings the rubber ring back, it is almost like a symbol of his conquest. In another example, which created equal social stature, Josh knocked out a bouncer after a disagreement. Conversely, potential mishaps are also constructed as glorious moments on holiday – even if they occur to the most unlikely group members. Here the Surreal Brummies recount some mishaps from the first night which, on recalling them in the focus group, seem to be as funny as when it happened:

Dan:	Josh pooed himself.
Will:	And went to the hospital.
Dan:	Oh really.
Marcus:	Hey, Josh, Josh, Josh, wanna tell us about the hospital story? [Josh looks over from his horizontal position. They all laugh as he reluctantly lumbers over step by step like a zombie]
Bob:	I'll tell you what's the best part of the story.
Marcus:	Tell him what he did the next day.
Will:	He [Stevie] stood there sober and pissed himself [all burst out laughing]. We were sitting in the children's playground [near el huevo] and he just pissed himself there [laughs out loud to himself].
Adrian:	I was there. Will's gonna tell the other story.
Dan:	Will's gonna tell the story? [smiling] OK, come on then, Will.
Will:	Well, it all started when we were in a nightclub and for some reason he [Josh] just passed out.
Dan:	He just fell over?
Marcus:	He only had a few, he was a bit tipsy but then he walked off, but after five minutes he'd come back.
Will:	And there were first aid people [Ibiza 24/7] but we carried him up, made sure he's alright, put him on a bench, called an ambulance and he pooed himself on the bench! [All laugh]
Dan:	So basically he was out for the count?
Marcus:	He had just come back and he had collapsed and he was throwing up all over the floor and I just carried him off the road to like a bench and the ambulance came over and stopped and he pooed himself just there [several laugh].
Dan:	Shit, I don't know, was it funny at the time?
All:	Yeah.
Dan:	[Surprised] It was funny at the time?
All:	Yeah, course mate.
Marcus:	I dragged him back and one of our mates walked in to get the money for the ambulance and they [the doctors] made it sound like it was really worrying.
Hugh:	And then he [Marcus] slapped him in his face and said 'Josh, you're shitting yourself'. [laughing]

Marcus: [in a serious tone] But they weren't really helpful at the hospital [Médico Galeno], they just put him on a bench, then chucked him on a bed, naked.
Dan: Really?
Will: [Continuing the serious tone] Yeah, it was like a waste of money, it was €590 for them to take him from the West End and leave him on a bed. And in the clinic next to Es Paradis [which is a matter of a few hundred metres] they wiped his arse and left him on a bed.

Interestingly, the conversation seems to take a more serious tone when, as they see it, the ambulance services fail to offer adequate care of their friend. They are even more critical when they find out that they are charged €590 for an ambulance to collect him, take him a few hundred metres to the clinic and leave him on a bed to sober up. This is valuable holiday money and there is no means to claim it back as they have no insurance. But, as we have seen, the private medical clinic on the seafront can charge what they like (Chapter 8). This was not the whole story of the first night either because in the process of deciding where group members were to sleep in the room, one was challenged to lick the other's testicles ... although it went a bit further in the end:

Dan: People have been saying 'What is done in Ibiza stays in Ibiza'.
Will: We don't want everyone to know about like the petty stuff, like them licking each other's balls and that.
Dan: 'Petty stuff like licking balls?'
Will: Yeah, the guy that pisses himself, Stevie, he licked his [Bob's] ball bag in one bar on West End [everyone bursts out with laughter]
Marcus: Licked his scrotum and then sucked it under his foreskin. [They all laugh]

Like the Southside Crew, the Surreal Brummies show a *habitus* already well moulded around the familiar concepts of the home NTE; in the way they have come to know that drinking and drugs help augment life's 'experiences' (Chapter 5). Similarly, they have some general idea of what is expected of them when they are in San Antonio (Chapter 6) which is reinforced when they arrive and reflected in the total bombardment of drink and sex offers (Chapter 8). With nothing certain back home, the spatial and situational shift promotes an ontology of 'seize the moment' regardless of what happens; it is seized – that's all that matters (Chapter 3). This flexible attitude allows for seemingly infinite diversity for deviance and risk into the realms of the bizarre and extreme – far beyond actions they would do at home and consequently more exaggerated. They search for *real* encounters from the unreal experience of the holiday resort and home life which collectively symbolise the material world (Chapter 3). The group, and different components of it as well as individuals, offer up feats of male endurance and stamina as well as homo-erotic, extreme and bizarre behaviour (Thurnell-Read, 2011). This is where the *jouissance* lies. In addition, the more varied their nights are – different combinations of group members, in different places, doing different things – the more assorted their collective experiences become which makes for enhanced 'drinking stories'. Where female groups tended to display more protectiveness over each other when they went out in groups and were more likely to feel guilty about transgression in hindsight (Thomas, 2005), male groups, by comparison almost encouraged deviance and risk-taking between themselves (Thurnell-Read, 2011). However, this didn't mean that the way in which pain and tragedy are celebrated was only a feature of the young male groups.

The pleasure in pain: From 'dream' to nightmare and back again

Gale: As we found out last night, I am dangerous when drunk. I walk into things, like last night I cut myself.
Dan: Was it serious?
Gale: [Giggling] I was sitting down and there was a metal bar above me, like, and I stood up to quick and cut all the way down my back.
[Shows me a large gash down her back]
Dan: Hmmm, yes you're lucky some of your bikini covers some of that – that's quite big.

Gale was on the West End the night before this interview when she cut pretty much one side of her back to the other. That night, she and her friend had pre-drunk a bottle of vodka and a bottle of peach schnapps, topped up with lemonade. They reflect that only after a few days into their holiday, they started to dislike the West End but reluctantly went again the night of this accident – after all, getting drunk on holiday is what is expected of them and what they subconsciously expect of themselves (Chapter 7). That night they saw a *'threesome in the toilet'* and said the bouncers locked them in the bar so they *'had to buy drinks'*: *'they really pressure you'* said Gale's friend (Chapter 8). Yet in the narrative it is clear that, despite their reservations about the West End, the way they were forced to buy drinks and the later accident, it was still a 'good night' – despite all the pain. Like the stories – from bizarre to extreme and heterosexual to homo-erotic – there is a pleasure in the pain (Lacan, 2006). Such is the determination to enjoy and have proof of that enjoyment that even agony and suffering have some sort of importance – perhaps less for themselves and instead more for how they can frame those incidents to others (Hall et al., 2008). This was the case for this group of young women who were interviewed in a café one morning after a particularly brutal night outside one of the Superclubs:

> It is around 11 a.m. A large party of girls in their late teens stagger home to the café where I eat a burger. I approach them for an interview and they reflect back on their time on holiday thus far where, on the West End, 'vodka was poured down their throats' and how they got *'free drinks'* just because they were female. Some even passed an ambulance near el huevo, in which was being loaded the body bag of a young woman. They later found out at their hotel that the dead woman was staying at their hotel as they saw her friend taking the phone call with the sad news. They had been to Eden and Es Paradis where at one foam party a guy had his hand in one of their tops and one had felt a hand up her skirt. Yet on this morning, most have sunglasses on; clearly it was a heavy night. But perhaps it was heavier than they would have expected. As I sit at their table where they order fried breakfasts, the conversation is dominated by the fact that only five hours ago, they said they were beaten by bouncers and the police. They start to recall the injustice of what happened yet there is an air of excitement about the story which seems to have stemmed from their drunken night in Pacha over a dispute over their alleged invasion of the VIP area without permission. It then seems that one was *'rugby tackled'* by a bouncer who also called them *'whores'*; the bouncers thinking they were *'high on drugs'*. Unfortunately, they started swearing at the bouncers and club manager. When the police were called, they continued their tirade at them and failed to cooperate and were handcuffed and arrested. One officer headlocked one and dragged her to the car and her knees and hips were cut and bruised; three others also had similar injuries [they show me]. In the end, they all laugh strangely at the affair as if it was something comical; two had even thought about potential Facebook

statuses to post when they were in the police car. One says 'Mine is going to be: "Welcome to fucking Ibiza ... already been molested and arrested"'. [Field notes]

Note how the initial narrative of injustice disappears, the evening becomes legendary as the tragedy is celebrated and the scars, cuts and bruises become the emblematic proof of the event. One could argue that my questioning prompted them to reminisce about the tragedy but, later in the interview, they confess to constructing particular Facebook statuses – even while the tragedy was 'live' as it were, it was already being savoured for the potential envy and attention it would create. This is therefore part of the rationale for transgression – even if it is unplanned: the more random, the more kudos and consequently, the more *real* (Chapter 3). *Capitalismo extremo* functions so efficiently by convincing these young people that, even despite the most brutal behaviours, these experiences are fun and enjoyable; that there is seemingly infinite pleasure in the pain and triumph in the tragedy. But even if these sorts of extreme events don't take place to the same extent among the cohort, *capitalismo extremo* still sublimely functions, unbeknown to those who think they are a cut above these behaviours.

A cut above the rest? Escaping chavdom ... but perhaps not *habitus*

I argued in Chapter 6 that some working-class British tourists in this book sought to appear as if they had one over on their counterparts on the West End – getting wasted each night. This is generally seen to be achieved by being/staying in Platja d'en Bossa over San Antonio; attending the more elite clubs like Pacha and Amnesia over Eden and Es Paradis or even working in Ibiza over being a tourist on the island. I suggested that there was no inherent difference as both groups were essentially getting intoxicated and that the only difference was the price they were paying for the privilege and this is another example of how *capitalismo extremo* functions, except this time it is more to the financial detriment of the British tourist. A good example of this is Sinead and her friends who took the short cut and came to Ibiza unprepared. This case study shows just how commercial the pull is to Ibiza as Sinead and her friends are generally unfamiliar with the music and instead only interested in the 'image' connected with being there (Chapter 6).

Sinead, and her five friends, all around 25–26 from working-class backgrounds in East London, booked a *'once in a lifetime'* ten-day package holiday to Ibiza; 'How can we not go while we are young?' her friend added. While two had been to Tenerife, this was only their first holiday abroad together (Chapter 6). Billed as the last *'naughty girls"* holiday before they started saving for their marriages, they were disappointed when they got to their hotel on San Antonio Bay to discover that the 'chavs were there drinking by the pool'. Equally, two of the five were disappointed that the group only managed to go clubbing twice as they went to see the big 'clubs'. It was certainly a particular image they were after when they went to Pacha. Without the spending power, they pre-drank wine and shots before the club as they didn't want to pay *'€15 for a drink in Pacha.'* When they went to another club, they couldn't remember who the DJ was but that they think it was called Space; even then it wasn't a good experience and the music was *'weird'* and didn't fit with their interests; plus they got *'pestered by some idiots'*.

Equally, Sinead didn't much like the West End – although two of her friends did because they like R&B. She said she could 'smell STIs, could smell teenage pregnancy, could smell urine and idiots'. Even though she had to slap a guy because he spanked her bum on the dance floor, she is still treated like an object (Chapter 8). Apparently, without *'wanting to'*, she says she got *'really drunk'* with her friends in one bar doing a deal for 15 vodka and cokes for five people. They lost each other because of festering holiday disputes and different music tastes. She had heard of Bora Bora so went there and liked the idea of going to Ushuaia, 'It looked like my kind of place' she said. It was the same for Blue Marlin where three of them spent a morning where there were British, American and Italian 'posers in polo shirts and white chinos, suede shoes and gelled-back hair'. It wasn't cheap though and a sunbed cost €100 to rent with a €150 per person minimum spend. Luckily that was quite easy as one round of drinks was €50 for three. In the end, they tried to leave early so as not to spend too much money but taxis were all booked up so had to stay longer, thereby spending more money. By the end of the holiday, Sinead had spent £1,000 but didn't count the £600 on shopping as spending (so £1,600 altogether). They all want to go back to see more clubs, staying in Platja d'en Bossa and maybe even renting a villa.

As we discussed, working-class Brits in this sample such as Sabrina, who want to seek a cut above the rest seek to do it in the 'better' Superclubs, did not quite have the capital to 'do it properly'. Some in her party even fall into debt and end up borrowing from each other until the end of the holiday. *Capitalismo extremo* not only catapults people into excess, deviance and risk but also gives them a glimpse of what they could be with further participation. More prominently is the social distancing from the West End – even though they spent a few nights going out there and getting drunk. To rationalise her overspend, she downgrades 'shopping' out of her holiday expenditure but she will be back because she has seen there are 'things to say one has done in Ibiza', and much like this next example, it could be the start of a long and beautiful love affair with the island because there are 'always new things to see'.

A cut above the rest? The experienced 'Ibiza goers'

In Space, I get talking to a group of posh-sounding British women who sip cocktails on one of the round tables. They are in their late 20s, early 30s. One used to work here while the other one has been every year for the last 12 years and it is these two who are chaperoning the other two for their first holiday in Ibiza (who immediately say they will return the next year). Although the experienced one knows Ibiza, 'there's always new things' she says. The same woman recalls her experiences over the years of seeing Brits roll out of Amnesia nightclub completely *'off their faces'*, some of whom were *'either killed or really hurt'* by night-time traffic. The woman who used to work here used to go out with a DJ so she saw 'Café Mambo, Judgement Sundays, Manumission and up the West End'. In fact, she never paid for drugs because he also dealt in cocaine and pills. These two women note how it is not as busy as it used to be on the island, one reflecting that 'It's the most iconic place and they'll be making more money out of you now, that's how it feels ... it used to be wild, like here [in Space] you can buy a bottle of water for €10!' Because two women know Ibiza well, they are not staying in the *'shabby part'* and have an apartment overlooking Café del Mar. It turns out this experience counts for something because they are on the guest lists for all the main clubs and are consequently paying nothing to get in, *'just drinks'*. One says:

Suzy: We're doing it very differently to how we did it obviously years ago, when I was 18 and 19, when your money runs out quickly and you don't eat and you can't afford to. Now you know you come with a lot of cash, you go out for nice dinners and yeah, we probably do it when we're sort of still getting older, but yeah, as you said, we're doing it very differently now.

They downplay their drug use and put it down to their age but then it transpires that all were on pills the previous night in Pacha. They claim to '*have the money and get off their tits*' but most nights, they have made dinner on the apartment balcony. They are also fortunate because the taxi queues are so long, they just call a contact they have who has an illegal taxi business (Chapter 8). As men become the topic of the discussion, they confess to a few kisses but the funnier things seem to be how a couple of them vomited on themselves while they were in bed one morning after a heavy night out.

In contrast to the last case study, this group of more mature women don't necessarily have more money but just have better contacts which makes the experience easier. Note how they are inadvertently supporting the informal economy as they have an established contact that can take them to the clubs and pick them up, no hassle (Chapter 8). However the interesting thing to note is that the behaviours they exhibit and describe in the more elite clubs like Pacha and Space are not far off what is taking place among the people supposedly below them in the social chain in the clubs like Es Paradis, Eden and even on the West End – they are also on pills and being sick on themselves, and even then this generates the same enjoyment from telling the story. So although people may think they have one over on others, that they have some superficially higher status, really they don't. This is the ideological power of *capitalismo extremo* because more status means spending more money but the tourist spaces, which are continually reinvented and commodified to accommodate new mediums of status stratification (Chapter 8), only create the motivation to come back again to claim that status or to say they have done something the 'other' hasn't.

The winners of *capitalismo extremo* (Earning a living off 'living the dream')

Donna: I'll tell you what else has changed. You know Café Mambo? You know they've got these plush seats and bars at the front. That didn't use to be there; it was literally just tables and chairs and then the rocks. Everyone used to sit on the rocks, but now they've made it so you can't enjoy Mambo as much unless you pay. It's separate, so you are separate [from the experience]. When I first came [in 2001], it was sand; they used to dump sand in there because it was a man-made beach. You used to have all these sunloungers and everyone would go in there, sit down and listen to music for free, take your own booze, cos you're at the beach, smoke weed or whatever and just listen to the DJs. Whereas now everything is commercialised and you must spend money. It's all a bit of a sham, like the opportunity, it feels like they are turning every opportunity into cash.

Donna here is describing yet another example of the commodification and commercialisation of the tourist spaces in Ibiza (Chapter 8). Café Mambo is another example of how marketing and business entrepreneurs with purely commercial interests have capitalised on the sunset 'experience'; aggressively marketing it as something which has 'to be done' in Ibiza – while at the same time, built around it a host of commercial ventures. For example, one can still buy the supermarket carry-outs and sit on the rocks watching the sunset, but it doesn't look very good because the really classy people have reserved tables in the bars which sit above the rocks and it is there they sit sipping and sharing €25 jugs of cocktail with their sunglasses on toasting the high life (Chapter 6). By comparison, below are the litter-ridden rocks where gather fag butts, empty bottles, cans and general rubbish. There is no sand, no sunloungers, no DJs (the initially underground sound of Swedish House Mafia recently moved their set to Ushuaia where they got a summer set): to enjoy the experience you need to sit in the private bars, get on a sunset cruise or private speed-boat, try laughing gas or the like – the whole experience is completely commodified.

It is the global corporations, marketing companies, commercial entrepreneurs and tourist operators behind ventures such as these which are the winners in *capitalismo extremo*. In the Superclubs, they charge high prices, turn a blind eye to drug use/dealing, have no accountability in the event of violence or the potentially serious consequences of excessive drug/alcohol use, have negligent security, are not bothered how many tickets they sell because it all means more money and seem to pay off the politicians and police to allow their activities to continue. Down the West End, new bars and clubs open all the time, the cheap deals continue to be aggressively marketed by PR workers under the illusion that they are 'living the dream', the prostitutes offer their services and get violated by the police and the tourists, the bouncers continue unregulated and the police can only make small inroads into the complexities of the drug and sex markets surrounding the West End – yet still the British tourists come, deal drugs, engage in deviance and risk behaviours and leave with all manner of individual, social and financial problems. This is because when *capitalismo extremo* is functioning efficiently, it is to earn profit at the expense of those participating in it and thereby ensuring its continuity. Just as described by Tiago in Chapter 4, it is 'cancerous', infectious; it destroys matter in its pathway and takes no prisoners.

The losers of *capitalismo extremo* ('Living the dream')

The losers of *capitalismo extremo* are the local Spanish businesses and the British tourists. Why is this? At the moment, anything to do with local, 'traditional' culture or remotely connected to what Ibiza was is being eradicated and replaced by a universal commercialism made up of global chains, brands and marketed symbols of the 'good life' (Chapter 8). We have seen how the local Spanish have suffered in this respect and many, if they aren't on the verge of closing, have had to either shut up shop or reluctantly join the bandwagon (see Chapter 4). Thus 'banal nationalism' is reinforced (Billig, 1995) in the resort context as other businesses fall into line with the dominant tourist group – the British. The other side of this, of course, is the way in which the British tourist loses out. As discussed, although many find the kind of things they do on holiday 'enjoyable', to me it seems difficult to ignore their impact on their individual, social, physical and emotional wellbeing. I have tried to show in this book how the ideology of *capitalismo extremo* – to 'seize the moment' and 'live the dream' – is in fact an illusionary layering which leaves people in a delusional state of mind that it was all worth it – even if they return home injured, with an STI, penniless and feeling existentially hollow, or worse off, with a friend in the morgue in Ibiza. In these examples, I show just how powerful the ideology of *capitalismo extremo* is because it is testament to the lengths people go to enjoy the 'good life' and/or 'live the dream'.

Checking in to a coma, then checking out for Carl Cox

I approach the end of an interview with a council worker who works in the drug field on the island. As the conversation naturally draws to a close, she finishes her coffee. As she starts to relax somewhat she recalls a story about a young British tourist whom one of her friends in Can Misses hospital treated last year [in 2011] in the emergency room. He had arrived from one of the Superclubs after heart failure from an overdose of ecstasy with alcohol. The staff, she says, were unsure whether he would survive and quite quickly he fell into a coma. However, 24 hours later, he miraculously regained consciousness. As he came round, he was advised to rest then return to the UK. To their surprise, he just insisted on asking what day and time it was. When the staff told him, he started to sweat and to get out of bed;

he checked himself out of the hospital because he was missing Carl Cox's opening night at Space.

Lost (in every aspect)

I leave Space but as I start to head back to the hotel I decide to deviate to see if I can catch a glimpse of the sunrise. I think I might be too late but among the people on the beach, I see one man just standing there with his eyes wide open, staring directly at the sun. At first, I stand from a distance hoping he will come out of his trance but in the five minutes I watch him, he barely blinks. He continues to gaze into the sun as if it has him hypnotised. When I approach to see if he is OK, he ignores me; his eyes bloodshot and strained. I look around to see if I can see anyone that may resemble someone who he may know but no one, apart from me, seems interested in him. I leave and go to get some breakfast. When I return, about 90 minutes later, he remains standing in the same place, gazing directly at the sun; this time his eyes look even redder and there look to be blisters forming in the corners of the eye. With some obvious concern, I shake him slightly but he doesn't respond. I leave and look for someone to inform, but the only person I can find is a policeman getting into his car who seems uninterested. [Field notes]

Still 'living the dream, Mark?' It seems so ...

Remember when I met Mark limping drunkenly across the marina with a bloodstained T-shirt and half an ear Mark in Chapter 8? Well, several days later, and in the same clothes, I see him lying on a concrete slab in the shade of the San Antonio morning sun:

As he stirs, he removes his dirty socks and I suggest I buy him breakfast if he can tell me how he came to be in this situation. We walk in, and while he goes to the toilet to wash, I buy him a coke and a croissant as he requested. We then sit down to talk. Although he has been to Ibiza before, he came back because he became absorbed by the place. Mark likes the music but initially came here to be a DJ in Ibiza. During the interview, it transpires that the damage to his ear was caused when a bouncer bit it off. However, for most of the three weeks he has spent here, he has been homeless because he spent his £2,000 – which he earned from doing medical tests in the UK – on a mixture of drugs and alcohol in Ushuaia and at casual worker parties. The money went in just three days and the only thing he can remember is buying his first drink in Ushuaia. Still, he tells me, he is 'living the dream'. [Field notes]

Mark is only 26, and after quitting his job in the UK several years ago, is now homeless in Ibiza, has no accommodation, no money, no return flight booked and had no idea where the consulate is.

Ambulance!

It is about 4 a.m. As we walk to the West End, we see a young woman on the opposite side of the road sitting on the kerb in the arms of her friend as if it is the end scene of a war film where the main character is trying to save his/her friend. The young woman cries, drifts between consciousness as she foams at the mouth, and as her head flops from side to side, her friend tries to save it from hitting the ground. Another stands over them, arms folded with her clutch bag, eagerly waiting for the ambulance. [Field notes]

I sit eating my burger and chomp away after a long day of fieldwork. A moment to myself, I think. Minutes later, at around 11.30 p.m., I hear some commotion at the end of the road, near the West End strip. I then start to hear ambulance sirens. I go down to see what it is all about and an ambulance starts to automatically marshal aside drunken Brits; they don't even seem to know it is a sizable vehicle which has just managed to drive through the droves of people. A medic gets out and goes over to a man who looks something between delirious and suicidal. Blood pours from his head as his friend tries to mop it up with his hands. Another wipes blood from his clothes. The man looks in disbelief as he keeps putting his hands on his bleeding head. He drunkenly sways and can't seem to summon any words to describe how he feels as his eyes blink in slow-motion at the continuing party around him. [Field notes]

Capitalismo extremo leaves these people out on a limb ... but there are ways to overcome this; there is a cure and that is to come back to Ibiza for more of the same!

Conclusion

The holiday occasion is a special time in which the absence of responsibilities and routine are celebrated (Briggs et al., 2011a), and this helps to loosen the boundaries of the home 'self'. This is further encouraged by the group dynamics in which many travel to Ibiza and mutually reaffirm the centrality of sex, ideological gender relations (Andrews, 2005), drunkenness and hedonism (Briggs et al., 2011b). The Southside Crew, like many British youth holidaying in San Antonio, drastically increase the use of illegal substances and engage in a rather different and broader range of risky behaviours. The holiday thus functions as an almost cathartic 'time out of time' (Bakhtin, 1984), and, in the absence of work, a quest for immediate desires prevails over even 'friendship', evident in the way in which groups dissipate on the West End and become at the mercy of the social context (Chapter 8). There is also some degree of taking pleasure in being 'wasteful' (Bataille, 1967), particularly evident in the way Jay and Nathan continued to party even though they could barely walk. Generally, the people in my book don't have the resources to 'party' like the elite regularly, but while on holiday they focus on the consumption and destruction of resources – most notably money, time and health – in ways that copy the expressive waste of the leisure class (Veblen, 1994) and suggest a desire to recreate the self in the image of the boundless celebrity consumer that is so central to popular culture (Hall et al., 2008, Chapters 3 and 5). Their *habitus*, their attitude towards the holiday (Chapter 5) and the expectations that they have constructed around the experience (Chapter 6) marry perfectly with a social context which has become increasingly commercialised and which has, as a consequence, developed a large informal economy intent on also draining money from the tourists (Chapter 8) regardless of the consequences. This is how efficiently *capitalismo extremo* functions: so that people spend and fail to realise why – even when it is to their total personal detriment. In fact, this 'good life' becomes immensely appealing – especially after some sort of evaluative confrontation has taken place that points to a redundant home life (Chapter 7). So it comes as no surprise that when the time comes to leave, most encounter a deep depression. The next chapter looks at what happens to the self in this back-home transition and how some construct a return to Ibiza as the cure to the malady.

10
Going Home ... Only to Come Back Out

I sit drinking with Nathan and Jay in their local pub. As we gulp our pints of beer, their Ibiza holiday becomes the discussion point:

Nathan: It was a year ago and I am just getting over it now. The last day, I really wanted to fucking stay. Didn't I say that that I was going to sleep on the beach and find a fucking job? I didn't want to go.
Jay: If I didn't have a family, I would be there every year. Fucking on it. It has to be done.

Over drinks, the eventful night is confirmed as something which has gone down in history: a tale for many years. Later in the discussion:

Nathan: Good night? And he was like yeah because I shit myself [laughs] and got sucked off.
[All laugh]
Dan: Is that the best or worst?
Nathan: The best! The worst was coming back [home]!

Introduction

In a conversation a year after their holiday in Ibiza, Nathan and Jay can clearly recall the 'great moments'. In doing so, they confirm the perception that what they experience in Ibiza – and for many Brits in this book and on holiday elsewhere – is a kind of constructed 'freedom'. For most, however, the feelings generated by the return to home life are sour and existentially penetrating, and this is evident in their admissions that they wanted to stay and that, despite everything – even the most bizarre moment of the holiday when Nathan was sucked off and shat himself – the feeling of coming home was more depressing. The perfect life it seems is one without these home pressures where they can engage in this kind of hedonism on a daily basis.

Why is it that they feel like this then? I think the return to home life produces a dualism in the self: that is, how, over the course of the holiday, a gradual self-deconstruction which produces a kind of emic and reflexive evaluation of the self takes place at the same time as an open-armed embrace is made of the life of the resort and the dream-like landscapes of the West End, the Superclubs and beach clubs (Chapter 7). This identity reversal reinforces the perceived 'shitness' of 'normal' life but also, at the same time, exaggerates the false happiness of the 'good life'; the *hyperconsumption* providing a bogus safety net, an imagined escapism towards security which easily stimulates the subjective need for *jouissance*. This often results in an internal existential tug-of-war as people like the Southside Crew end up not knowing what is real and what is false; they lose their ontological bearings. Home life is laid naked before them and there may even be some small realisation that the dream-like essence of what they participated in was actually a nothingness. What they are doing is reflexively looking 'below life' and it feels like looking deep into a desolate and personal abyss; a psychological no man's land where the false is real and the real is false.

I want to argue that young Brits deal with these feelings by: (a) going out when they get home to quickly revive the memories and home excesses as part of their *being*; (b) to start planning/saving for the next holiday next year, potentially in Ibiza; (c) impulsively going out to Ibiza the same summer or even fail to return and stay out there. Their return to Ibiza is also buttressed by a significant amount of marketing around the 'reunion parties', the release of Ibiza club compilations which stimulate memories of the 'good times' and the continued discourses which revolve around leisure and online time (in the pub, on Facebook) about the perceived magic of the island. We are also therefore talking about a *commodification of nostalgia* and this also assists with the personal impetus to consider returning. Let us firstly look at how the Southside Crew's holiday ended.

The beginning of the end: Taking the party to another level

As the holiday approaches its end, many British groups in my cohort tend to turn up their partying to the same sorts of levels as the first night. Those on a short break to Ibiza are more likely to power through in the few days they have while others, who book package deals for a week or more, often have a lull in the *hyperconsumption* (Chapter 7). This seems to be the way to mark the 'time out' and, once again, offers more of a personal rationalisation to transgress into the bizarre and extreme; the general attitude being 'It's the last night, better make the most of it'. When I see the Southside Crew two days later after their first night, they seem visibly upset that the short holiday they have – this window in which they can seize this amazing reality – will shortly close. It is about midday when I see them on the penultimate day. What follows is something between a reflection on what has happened and also reaffirmation that the party needs to be taken to another level.

Approaching the last night: The morning before the night after

Having knocked on their empty hotel rooms, I hunt around the beach area where I see Jay and Nathan sitting on the small concrete wall which divides the dusty road and the cigarette-butted beach. I walk over and catch up on events. They all nurse massive hangovers and have hardly eaten for two days. Still, the party must continue right to the end:

Dan: [To Nathan and Jay] Are you alright?
 [They shake their heads with some slow difficulty]
Dan: Aren't you off home tomorrow? Tomorrow night?
Paulie: [In very gruff voice and opens his eyes to speak] Yeah.
Dan: Shit.
Jay: We have tonight to party hard, mate.

The last night beckons. Between recounting the night we met, they discuss what they did last night: the day after I met them. Jay and Nathan, it seems, were drinking until Jay *'passed out'* while Nathan was off his face on pills and ended up in Es Paradis; he was 'fucking tripping and ended up sitting, shaking my head all night' he said. Once again, he ended up by himself. It seems he had bought them from Simon who had come to Ibiza to do some drug dealing. They continue recalling what happened last night; it seems Paulie took an E and went on the bungee jump ride; he says he will put it on YouTube and Facebook when he gets back. The laughs pick up again for a short period but there is then a silence and Simon starts playing with the fag butts in the sand. The imminence of home is on the horizon and the self-reflection seems to continue as they go between playing with the sand and looking out across the bay. In an effort to cheer people up, Simon suggests going on a speed-boat and rubber-ring ride for €10 but only Paulie seems willing. They both leave after searching their trousers and swimming trunks for enough coins to pay. Nathan sits on the wall and cracks open another beer: 'Right, I am fucking going to get on this' and points at his beer.

The conversation drifts back to last night, as Nathan admits to buying laughing gas *'off some bird on the West End'* (Chapter 8). Drawn back to the same bar because of *'a mate'*, they claim it was because of the alcohol deals – yet most of the deals are the same. Then, after learning more about how the deals work, they reflect on the night they shared with me, saying they were *'mugged off'* on the Clubland deal because they were *'learning the place'* yet they had been to San Antonio before. The conversation stops and starts between lengthier reflections, but my attention keeps turning to Nathan who sits on the wall in his swimming trunks, a broken straw hat and dark sunglasses. Every time he takes small sips of beer, his hands visibly shake. He then pokes fun at Marky saying the only reason he doesn't do pills is because he *'goes Down's syndrome face'* and makes an impression. Yet between this banter there is a sense that on the journey which is the holiday, one needs to outdo the other; that there is a competitive element to it, and with the last night on the horizon, it seems that things will be taken to another level:

Dan: Do you feel like it's a competition between you?
Nathan: Yeah, definitely.
Dan: So who is winning and losing?
[Long pause]
Marky: Don't know mate, who knows. Ask me in an hour when I am lashed up [laughs].
Dan: Survival.
Nathan: Survival of the fittest.
Marky: Basically, you come out here to get fucking minging. You got no one stopping ya. You haven't got your missus going 'fucking behave', you know what I mean? You can do what you want and that is why I am here – to do what I want and get away with what I do.
Dan: Yeah, yeah, yeah.
Marky: No one is here to tell us off, we can do what the fuck we like.

[A plumpish girl walks past in a thong bikini and the guys cringe]

The competition is not only to establish who can consume the most but also who can potentially generate the most bizarre story through playful deviance and risk-taking (Chapter 9). Because it is the penultimate day, they feel a pressure to turn up the drinking and partying before the imminent crash of going home, Marky saying 'I'm staying here for a while, mate [laughs]. I am not going out early tonight, mate. Go out at 10 p.m. mate, soak the sun up and then abuse it. That's my day, anyway.' Marky knows his night could end prematurely if he drinks too much during the day, but this doesn't stop him from drinking on the beach.

Last day: The morning after the night before

The next day, I meet the Southside Crew for the last time by the beach. It is early afternoon and there is a sombre atmosphere. As a gesture of thanks, I buy them all cold cans of beer. They seem to appreciate this saying it is 'better than the free shit the hotel gives out'. Simon sits once again, playing with the sand while Paulie leans back against the concrete wall and looks at the sky seemingly regretful. Jay and Nathan have periodic glances out across the bay while Marky lies on his front and makes a limp sandcastle out of the dry sand. They have made acquaintance with two girls who sit next to them; one of whom Jay had sex with last night. Well, he did want to do what he wanted and get away with it. They still haven't eaten properly it seems: conscious of body image on one hand and not being able to face any food on the other.

The pressure for a 'last night of madness' seems to have equalled the level of deviance and risk from the first night; they seem to have taken things a step further in the knowledge that the horizon of this dream-like reality now sits in front of them at its conclusion. Unlike the first night, their recollection is patchy and only spurred by flashbacks. Marky and Paulie went off together but Paulie later left to have sex with a girl, even though he was moralising about cheating in the focus group when I first met them. Jay and Nathan continued the party and mixed drugs with alcohol on the West End. As they recall other events from their last night, we are almost constantly approached by African men who offer us sunglasses and/or drugs:

Nathan: When you got the drugs and put them in my cocktail jug and that. It went all green and shit.

Marky:	I can't remember fuck all, mate.
Dan:	OK, I am lost already.
Jay:	Can't remember that. It was in that English pub. Shit, I do remember now. I put all the drinks together, thousands of them in one.
Dan:	I am trying to remember when I saw you.
Paulie:	[Confused and losing touch with what has happened] Are you sure it was yesterday? [Another African man comes by to sell sunglasses and drugs. He firstly makes a smoking gesture to indicate drugs and then flashes his armful of sunglasses at me even though I am wearing a pair.]
Dan:	I'm OK, thanks.
African man:	More sunglasses?
Dan:	It's OK, I've only got two eyes. [They laugh]

The construction of events is slow and hampered by the fact that the five days have started to blur. Paulie cannot even remember when he last saw me – even though it was the day before our conversation. It then transpires that Jay and Nathan went to a strip club where they spent hundreds of euros because 'We had to get rid of our money, SPEND it,' says Jay. Nathan got carried away somewhat and 'was licking her arsehole and that'; she also licked his and *'pissed on him'* while Jay by comparison had three lapdances. While most got in between 6 a.m. and 7 a.m., Jay went back to the hotel early at 2 a.m. because he was *'too minging'*. But as we turn to their journey home, it seems they have a plan to counter the depression; they are to 'go straight out on Southside when they get back'; after all, it will be a Friday night. This excites them as Jay adds 'So it's Friday night and we still have Saturday and Sunday!'

The Southside Crew are aware their holiday is soon to end so turn up the *hyperconsumption* on the last night for one final push. However, to counter their quite reflective states, they seek to continue the party back home which will then cushion the post-holiday blues. In particular, Paulie's conversion from 'moral father' to 'cheating partner' exemplifies the pliability of identity in the holiday context but also the way in which Jay and Nathan rampantly indulge in spending hundreds of euros in the strip club: signing out with a bang by indulging in sexual fetishes. However, in the moments between these discussions, there is some reflection and the self starts to realise home life is drawing closer. In this process, reflexivity comes into play and prompts the self to take advantage of the holiday reality; to squeeze out as much from it as possible regardless of how difficult it may be and the fragility of their state. However, the triggering of this reflexive process also starts to fuse with the gradual self-deconstruction which has also been taking place over the holiday period. Suddenly, the banality of home life looks even more shit against the euphoria of the holiday moment and the limitless fun of the resort – and this is the power of the ideology of Ibiza's image (Chapter 7). This is because many of these young Brits, having convinced themselves that home life is shit and having embraced Ibiza's ideology, gain a dangerous perspective on themselves and their life situations; they end up staring directly into a personal existential abyss of 'below life' when they leave. This is what I mean.

Confronting 'below life' and the *reflexive double door*

Jackie: I was so sad [when I came back] I actually cried. I stayed at my mum's because I couldn't travel to my home town straight away. I was just in a proper state; I didn't want to leave, it was really sad to be home and I didn't want to go back to normal routine. I was mainly thinking 'I wish I could just be like the neighbours' [the ones who were living out there dealing drugs] and live out there and do that lifestyle for the rest of the summer.

I want to suggest that Jackie here is confronting 'below life'. It is when life is reflected on thus that a new ontological channel is prized open: one in which the self obtains a degree of objectivity on itself and its participation in the social system. Of course people don't describe it as I have thus but the way in which some try to articulate this sudden crisis in personal identity suggests this is another important transition which takes place during the holiday period. Looking 'below life' is a realisation of socio-structural position and prompts spurious evaluations that home life is crap; often even more crap than they thought before they came on holiday. This is because while the self has loosened itself from the home anchoring, at the same time, the week of euphoria has distorted that very same existence; on reflection these young Brits are left wondering what is the 'real' reality. This stimulates ugly thoughts and feelings, as well as producing insecurity and self-doubt. In this next excerpt, the concept of 'below life' arises through the data and I feel adequately describes what many Brits feel when they return from holiday. Let's rejoin Liam and Graham's conversation from Chapter 6 as they reflect on their *holiday career* and where it may be meandering:

Liam: Obviously when we were 18, we were like going to places like Zante, drink, drink, drink, birds, birds, birds, competition, competition, competition, but Ibiza is like chilled out, mellow.

Dan: So when does it stop? Do you think you will keep going back?

Liam: That's what I always said. First time I come back from Ibiza, I said to myself I'm not going nowhere else. Maybe Las Vegas where there is something in the desert but that is a one-off week event, like fucking on it. I would do Ibiza, twice a year for the rest of my life. Easy.

Dan: Why?

Liam: Love it too much. Music's good. Clubs are good. Now I've been a couple of times, I know what's going on.

Dan: But you know what's going on in Malia and Zante ...

Liam: Yeah but that's a shithole, mate.

Dan: So why did you go in the first place?

Liam: The only reason I wouldn't go to Ibiza was if I didn't have the money. If I had the money at 17, I would be straight in Ibiza. It is so good, everyone wants to go. Tell me one person who come back from Ibiza and didn't enjoy it.

Dan: Me.

Liam: What's wrong with you mate? It's got the best clubs, the best music.

Graham: [Putting his arm around me and trying to sell me a vision with his hand gestures] Think about it, mate. You are lying on the beach, clear blue skies, sun beatin' down on you, drinking alcohol and loads of fit birds walking around. You can go in the sea if you want, go in the pool if you want. Everyone is doing what they want. That is what is 'happy'.

The interesting thing here is when I place myself as an outsider to the 'Ibiza dream', I am corrected as if there is something wrong with me; perhaps this is why many people 'say it was good' because they don't want to look like someone who didn't enjoy it or who hasn't experienced Ibiza. In addition, the ideology behind Ibiza as a place for these activities (Chapter 7) skews the reality of home positions and this is evident in the identity shifts from home to holiday and vice versa. What I think we are therefore seeing here is a temporary glitch in the self as the end of the holiday approaches; a mishmash moment when the ideology almost convinces, and in some cases wins over, the self to reconsider everything at home. This is when 'below life' is seen. Look at what Liam describes here:

Dan: When you go home, how do you feel?
Liam: Fucking shit.
Graham: Shit because you know what you're going back to.
Dan: What are you going back to?
Liam: Like if you are in Ibiza for more than five days, [tuts and sighs] you start looking at your own life.
Dan: Looking at what?
Liam: Literally, you sit there and you think, 'What have you got to go back to?' You start looking below life.
Dan: It's a nice concept.
Liam: What I mean is like look at everything, your job, your girl, everything. What you are doing. It makes you think more about these things, like below life.
Graham: No one feels good coming back to normal life.
Dan: What is normal life?
Graham: Working and it's fucking shit. Everyone I work with are dickheads.
Liam: My job is boring as fuck.
Graham: When I get to 60 I don't want to say 'Oh, I worked at lot'. I want to say I have seen and done a lot.
Liam: A lot of people go Ibiza and start looking at their own life to see if it can, like if they can make it better. But they can't. They can't make it better. This is what kills me. When I can see my life like that.

Looking into this personal abyss which is 'below life' does several things. It is first a general evaluation of the self: where he/she is, who he/she is with, what he/she is doing. This strays into a realisation of socio-structural position and generates feelings of powerlessness because of that position. However, the same process also produces a kind of *reflexive double door*. As well as evaluating back on life, people also reflect forward and almost see their destiny – which isn't pretty either because most feel they can't do a great deal about it. They have all these dreams but they can't be realised since most not only have fairly uncertain futures but also precarious work commitments. Perhaps for a minute, a few may even see beyond the ideological blinkers which have narrowed their pathway thus far and can't quite understand why they were so wasteful with money and why they did what they did on holiday. The potential reality of the situation is that there isn't much to them – there is little to their *being* – and when people are exposed to such a hollow core, because what they have built themselves around is purely material and has little substance (Chapter 3), they end up staring into a personal abyss. How then can this possibly be filled? By 'getting on it' when they get home and/or coming back out again, of course! And this is, for some, enough for them to rebook trips as soon as they get back or, better yet, fail to return home. Take this public discussion forum on Facebook which promotes a discourse about Ibiza 2013. For example, on 8 October 2012, Becky from Chapter 5, posted an image titled 'I'm going to Ibiza in 2013 … and nothing is going to stop me'. By that day, it had generated 4,003 likes, 358 shares and 179 comments. Some comments ebb between '*defo*', '*too fucking tru, you just know this shit*' and '*every year Ibiza*'. Between the comments, such as '*already booked*' and '*about to book*' were further signs that the dream is alive and kicking:

David: I'm moving there and never coming back.
Jordi: Just arrived home today and waiting for 2013.
Jake: Hell fucking yeah!! I've packed already!
Johnny: Even if I need to imagine it, I'll be there!
Craig: But I'll be 46, is that too old? Is it fuck ☺
Ruma: Go for it, follow your dreams!
Nikki: Ushuaia, June 2013, hen weekend! Can't bloody wait!

Another way to counter the existential void is perhaps even more extreme: fail to return home from Ibiza, and/or return the year after to work a season and 'live the dream'.

Tim and the Kent youth: The road to 'living the dream'

The chapter thus far has argued that returning to Ibiza, or constructing the rationale to attend a mega-event or another holiday, comes as the self starts to come to terms with how enjoying the 'good life' for a short period, on one hand, depixilates their home existence – what they do, who they are with, where they are going – and, at the same time, sheds a bright spotlight on it thereby generating an existential void. Another way out of this position is to return as soon as possible or even 'live the dream': to work and live in Ibiza. This next case study, based on data gathered over two years, draws together all these elements to show exactly how this process takes place. It is the story of one young man called Tim and how he graduated through various other holidays, came to Ibiza, got hooked on the lifestyle attached to it and then decided to 'live the dream' in the summer of 2012.

The Kent youth

As some researchers and I walk along the beach in San Antonio, Ibiza, we get talking with three young working-class men (Paul, George and Simon) from a group of nine from Kent, UK. During the focus group and subsequent afternoon, they manage to drink around 12 pints of beer each. At home, as a group, they drink just twice weekly. They confess to being cautious about how much they drink (even though it does get out of control they say) and where they take drugs because of their local image. They regularly get into fights in their home town on weekends but not with each other; only with *'randoms who are out to cause problems'*. At home they attend trance events, some of which are reunion parties from Ibiza. Only one of the nine has a girlfriend. He says he is *'true to her'* but the rest laugh when they recount how they all paid for sex in a brothel the night before last. At home, they spend around £100 on a night out, depending on the destination and occasion. They have been away in a group to Faliraki on the Greek island of Rhodes, Barcelona in Spain and, most recently, to Amsterdam in the Netherlands. In Barcelona, one of the group hospitalised another in the same party by battering him with a beer bottle. In Amsterdam, they say they took 250 pills between four of them in a weekend. Mostly they are attracted to these destinations for dance music festivals which they attend at home, although some in the group just prefer getting *'gattered'* (drunk).

San Antonio

They arrived on their holiday in Ibiza drunk and high on pills, and with only one hour's sleep after arrival, went drinking in bars and then clubbing. Each of them has brought £1,500 to spend over one week – this is after they have paid for flights and accommodation. A typical day and night in San Antonio sees them get up at around midday, swim, eat breakfast, drink beer throughout the afternoon, swim, move on to cocktails and shots in the evening, taking various drugs. In fact, their extremely slim bodies give me a clue that breakfast is their only meal of the day when on holiday. After a few days in San Antonio, they have already spent €300 on pills alone and been in various fights with each other. Nights out are a mix of spirits, mixers, shots, cocaine and ecstasy often followed by early mornings, continuing the party with spirits and ketamine. They booked a booze cruise but missed it so booked another; a techno boat where it was *'binge drinking on another level'* says Tim, who estimated he had 12 double vodka and cokes in an hour. One of his friends, George, lost €90, blamed the others and got into a fight. George got even more violent when he discovered that he couldn't get into his room because two others had *'got birds in the room'*. Another, Ryan, got in a fight with someone on the West End, ran away crying, thinking he had killed him, and then threatened to jump off the hotel balcony. Despite the pleasure in the pain (Chapter 9), still they would drop everything to do these things every day:

Paul: I would give it up [work], yeah. Six months. I don't want to work over here, I just want to get smashed. Work in the winter [as an electrician] and come out here. It's hard to save money because we all meet on weekends and it costs a lot of money. Even our holiday, we saved up a lot and stopped using a lot of cocaine.

Tim: I took a loan out to come on holiday. I would live here, smash the clubs every night.

The imminent homecoming

After following their holiday antics over the week, their time in San Antonio starts to draw to a close:

> I walk down past the pool area at about 7 p.m. where it is pretty empty. Still the place is littered with plastic cups and there is a whiff of vomit coming from somewhere but I put it down to 'the sea air' smell. It sort of reminds me of a Great Yarmouth pub but with good weather. I walk down to Tim and Paul who are sitting on the beach. A couple of muscle-bound Italians play athletically with a tennis ball, throwing themselves in the water so they land in it, only to get up and flex all their muscles at once before diving around again like playful Adonis men. I go for a quick swim, and although the water looks temptingly clear from the outset, once inside, I find the glamorous fish instead weave in and out of empty beer cans and bottles. When I get out, I walk over to Tim and Paul.

Tim has his hat on and sunglasses on upside down. Paul just sits there with his sunglasses on. I sit and we look out over the sun setting, and although it is still hot there is a sombre atmosphere – probably because they are going home in two days. They don't want to go home and want to continue with the hedonistic non-stop celebrations (which surely must be having a strain on their health – we later learn that on Saturday after we left one of their group was rushed to hospital because a vein burst in his testicle and went black).There is some silence and I ask Paul how and why he does this [engage in such hedonism]. He says 'I just love it, I fucking love it. The music just goes through me. I could do this every day of my life.' [Field notes]

The reflexive process has begun for Tim and Paul. The holiday has allowed for some objectivity on home life and, over the course of their time away, the 'good life' and all the trimmings have certainly made some of them think twice about what they do back home. In the UK, existence looks so grey and ordinary yet in Ibiza it is colourful and extraordinary; as we have seen in what they do on holiday, the nights are packed with action, music, adventure, transgression, drama, *hyper-consumption* – this is the life they want, this is the 'dream'.

'Living the dream': Tim's adventures

Most of the Kent youth returned to Ibiza the same year (2010), the following year (2011) and, recently this year (2012). However, of all nine, it was Tim who decided to pursue the 'dream' and live in Ibiza. Tim is 24 years old and generally has uncertain manual labour contracts (Chapters 3 and 5). His teenage years were predominantly spent DJing, clubbing and raving around Kent with his schoolfriends who also got into the house music scene. His first holiday abroad was to the south of Spain in 2008 with his family; however, this was followed the same summer with his first 'lads' holiday' in southern Spain. Most nights were spent drinking heavily, taking Es in the local bars and clubs and ended up in drunken games at the end of the early morning. Between holidays, his weekends are spent clubbing and raving and between all this, he regularly attends festivals, private DJ venues and illegal raves. At some of these events, he has spent whole weekends raving and taking Es. His friend also runs a pirate radio station where he also appears on occasions. In 2009, the following year in Ibiza, they took a package holiday, drinking on the plane and not stopping throughout the week they were there. The highlight of their holiday was a YouTube post of a videoclip of them returning from a night out and caking one friend from head to toe in a crude mix of toothpaste, shaving foam, suncream and deodorant.

It was also in this year that his DJing started to gain some recognition and he started to appear in local venues with his friend, Paul, across the Kent area. His intentions to return to Ibiza are posted on Facebook numerous times late in 2009 and into early 2010. One such status is 'I can't wait to go to fucking Ibiza' and it transpires he is to travel out in May 2010 for the opening parties. There then follow status updates which reaffirm his commitment to techno music and all the big club DJs such as Carl Cox and the big club events such as Circo Loco. Apart from the clubs, he also boards *'techno boats'* which cruise around all day.

It was at the end of July 2010 that I met him and the other Kent youth in San Antonio. However, that same summer of 2010, he returned twice after I met him, principally for the closing parties in September. So in total, he went to Ibiza four times that year. When he returns home, his Facebook posts mostly concern clubbing and techno house parties. For example, early in 2011 he states 'Fabric is going to get battered on Saturday night'. Some of the parties he attends in the UK are reunion parties while others are familiar to the Ibiza circuit such as Circo Loco, which is as much found in the UK as it is in Ibiza. Perhaps without the same capital, he can only manage to visit Ibiza twice in 2011; once midsummer *'for a holiday'* and again for the closing parties.

In February 2012, he moves to Barcelona to *'get nutted'*. He stays there throughout March with his new girlfriend. They then fly to Platja d'en Bossa in Ibiza late in April where he looks for work in the area – much to the envy of all his 500 Facebook friends back home in the UK who throw comments at him for his bravery in 'living the dream'. Without work, he manages to rent a flat but it is with 20 people and there is one shower. However, by mid May he still hasn't found work. When his friends periodically visit him, he then has to accommodate them in this cramped worker apartment. Finally, late in May, he manages to get a job as a PR worker on a booze cruise in San Antonio. Much to his delight, his new PR friends seem to be well connected and he starts to attend some private parties in exclusive villas: *'living the dream life'* is one status following a mammoth private party which lasted a whole weekend. When he starts to attend the opening parties in late May, he struggles to hold onto the job and loses it after failing to turn up a few times to work. However, he gets some more luck when he is asked to

DJ in a small bar on the West End which leads to him playing a few sets on a booze cruise. This lands him some work on another booze cruise where he tries to persuade them he could DJ but can only get a job as a worker. For most of the summer, he continues to work for *Lost in Ibiza Sunset Boat Party*, in which naked women dance around on a boat to dance/house/techno music and the general aim is to get high/drunk. Tim confesses, however, to not remembering what happens most nights and it is perhaps unsurprising to see him in photos climbing out and balancing on balconies and climbing between apartments. When the summer finishes, he is back to his former home life; his Facebook status in capitals in November 2012 *'I NEED A JOB CAN ANYONE HELP ME! X'*.

Tim's story exemplifies many in the sample because it shows how the ideological power of the 'good life', on holiday in places like Ibiza, easily swallows up the reality of the home life. Tim is also a perfect example of how, throughout his *holiday career* (Chapter 6), he has come to learn about what he should do on holiday, where is the best place to go and so, once there, should start to seek more 'status' on arriving there (Chapter 7). The ideology of Ibiza is so firmly embedded in his sense of self that he goes there for holidays as well as clubbing experiences; for the last three years, every year he has been at least twice to the island. His Facebook statuses and photos contribute to a discourse about Ibiza, the clubs and the madness as much as it creates a social envy for what he is doing; one which may also potentially draw others into thinking that if he is 'living the dream', why can't they? And although his friends may not have 'lived the dream', they still return to Ibiza each year like the millions of other Brits and this is also testament to Ibiza's ideological pulling power. Of interest also are the reunion parties and the club nights which also feature in Ibiza: this helps to keep the notion of Ibiza alive through the winter months. The ways this takes place I think are reflected in a *commodification of nostalgia*.

The commodification of nostalgia

Ibiza is not really about hippies any more. Nor is it much about the freedom of rave and house music. What may have started off as 'something beautiful' in the name of 'unity and togetherness' by the four pioneers of the Balearic beat – Paul Oakenfold, Pete Tong, Danny Rampling, Nicky Holloway – is now a global commercial and brand industry which has become embedded in many of my participants' lifestyles and life outlooks. When these DJs opened up the Balearic beat to London clubs in the late 1980s, here began this commercialisation process because this was the first attempt to recreate Ibiza's music, image and atmosphere in the UK. At the time, it was said that rave united people beyond margins of race, class and culture in a unification of devotion to house music; rave was about 'togetherness' and 'dancing on drugs'.

But now the commercial landscape has changed. All the main Superclubs have their own CD collections, brands, shops, clothes, styles, and marketing entrepreneurs. Some in my sample are more familiar with the brand than the DJs, and the club than the music which speaks volumes about what has happened to the iconic 'Balearic beat' and the image of 'unity' and 'togetherness' which was Ibiza. This mass commercialisation has mainstreamised the once underground dance/house scene; the was-once special something is now a very unspecial everything. This has not only resulted in the branding of the clubs, and in some cases, the DJs but the widespread commercial realisation to big music names and large-scale festivals. This branding alone is found all over the island in the clubs' own stores, in markets, on posters, banners, airport merchandise, and, as a consequence, can be found in some form on most people's sunglasses, T-shirts, moped helmets or even baby outfits (Chapter 4). What happens as a result is that people become absorbed by Ibiza's ideology, its name, rather than anything else:

> On a booze cruise, I get talking to two young women about what they've done so far on their holiday. They swig from their secret stash of cheap vodka as they confess they too were in Space last night; the only thing is, they don't know who Carl Cox is even though one bought a 'Space' T-shirt. Perhaps it is no surprise they left at 2.30 a.m. [Field notes]

These are the people who have taken the short cut to Ibiza and who have been sucked in by the ideology, image and the status which comes with it (Chapters 6, 7 and 9). This exodus cannot simply be blamed on people hearing about Ibiza because the once eclectic music mixes of DJ Alfredo are now available to the masses as well as various cultural events and festivals with other types of music. BBC Radio 1 play a weekend in Ibiza which now runs every August and then there are the DJs whom some of my participants refer to as 'legendary', the more experienced Ibiza goers referring to mainstream DJs like David Guetta as *'puppets on a deck'* or even B or even C-list celebrities like Mark Wright from *TOWIE* who *'press play'* on the DJ decks (Chapter 5). In this way, the market opens up its welcome arms to receive anyone who is reasonably famous to play while at the same time invites with pleasure anyone willing to 'pay for the experience'.

Global corporations and marketing entrepreneurs, who put minimal amounts into local amenities and care little for the welfare of the island, are happy to organise massive festivals if it means that people will come and spend money. Take the recent Ibiza 123 Festival which ran for three days in the summer of 2012 (much against the persistent complaints from local residents and businesses). Three days' access cost €170 and this didn't include food or drinks. Sponsored by Bacardi and Monster Energy Drink, it was designed for a broad audience by the very nature of its varied guest list; a mixture of famous rock and dance artists such as Sting, Elton John, Lenny Kravitz, Luciano, Tinie Tempah, Tiesto, David Guetta and even some of the 'old skool' Ibiza crowd for the more seasoned Ibiza goer such as Pete Tong and Fatboy Slim. Something for everyone and this, to some degree, assists in the process of *time folding* (Chapter 6); the way in which Ibiza is constructed with the 'youth' experience as it is with 'nostalgia' not only resulting in the generational overlapping of its tourist cohort but also in their self-constructions to relive those moments.

But let's say you have been to Ibiza a number of times. For you, it has become a place to where you can relate good memories, then you will come back to it to relive those times. And if that place is marketed to you, as a place where you can rekindle those memories, then you'll come back to it. We see a similar marketing process in the way old skool tracks, albums and compilations are marketed in the UK. This represents a *commodification of nostalgia* – a way of persuading people that the good old times can and should be rekindled and this can only be done by at least buying the music or at most going to Ibiza again to recreate those moments. It is this which Liam was describing when he said he *'loved Ibiza'*; this is because he has come to believe that the 'good times' are in Ibiza and should be rekindled as often as possible.

So to consider these thoughts and decisions to return to Ibiza as autonomous to influence is foolish. People may think these things but I want to suggest that a *commodification of nostalgia*, cleverly guided by media, global corporations, marketing entrepreneurs and the music industry, assists in these ambitions to return. Just look at Hedkandi events at Pacha London which take place in the winter months just so Ibiza and its moments can be relived during the greyness of the cold season. This is, of course, all supported in the kind of discourses which circulate around Ibiza on online forums, casual worker websites, and social media like Facebook; consider Becky in Chapter 5. But further evidence of how nostalgia has been commodified is through the compilation CDs of the Superclub DJs, the reunion parties and the festivals and mega-events at which the DJs attend. So the party continues or can continue in the UK and memories can be relived. This is even reflected on the West End where there are Irish pubs, 70s and 80s clubs, hip-hop and soul clubs – something for everyone to turn back time and relive those precious moments of youth. But really all this acts to keep current the idea of Ibiza and, more pressingly, the thoughts to return to have fun and spend money.

Conclusion

Many people in my sample don't want the holiday to end: it's such an exciting time, one which does not really parallel with their home lives and even what they do at the weekends in the clubs and bars. This means that as much of it must be seized as possible and this is done to some extent by engaging in forms of *hyperconsumption*. This gives them something to take back which acts as the receipt for what they have purchased. Because the holiday is a finite period, as much needs to be made of it as possible, even if it means expending every ounce of energy, is at the expense of health or to the detriment of the bank account – or even social relations. Nothing can or must get in the way of making the most of this moment. This is often why people power through and continue to engage in *hyperconsumption* and the last night is another example of this. With the advent of the end of the holiday, however, the reality – which is the resort and all its euphoria – makes home existence look even more overcast and in facing the prospect of this, many end up looking back and forward on themselves through the *reflexive double door*. 'Below life' reflections are stimulated which are existentially disturbing moments because they reinforce in many ways the dreams which people in this book can't realise. While a few may make self-promises to navigate out of what they consider to be oppressive home territory, for most it is instead the catalyst for further participation in what they have just done – a return to the weekend excessive consumption, booking another holiday or better yet 'live the dream' next time round because no one wants to concede nor accept what they have just participated in: a nothingness which was pointless. They may say they are making these 'return-to-Ibiza' decisions at their free will but I have argued that there are numerous industries and institutions which have commodified nostalgia in an effort to make people think that a 'return to the good times' is necessary. In the penultimate chapter, I would like to discuss my findings within the relevant theoretical and empirical literature after making a succinct conclusion.

11
Discussion and Conclusion

Introduction

In this penultimate chapter, I will try and tie together my empirical findings with the theoretical context for my study. I have attempted to further the field of social knowledge as well as draw attention to the current social system and the way in which it assists in the reproduction of serious social and individual problems. I'm not moralising anything here; there is nothing wrong with going on holiday, drinking, and generally taking the stress out of life back home. However, I hope the book shows the darker side to this process, and especially for a certain cohort of working-class British tourists, how the holiday can easily become something unexpectedly perverse and dangerous. My thoughts on what I have tried to do begin this chapter and are followed by some methodological reflections. There then follows my conclusion followed by some thematic and theoretical discussions before I offer some recommendations to address these problems.

Aims of the book

The aims of the book were to address a significant gap in knowledge regarding the subjective meanings these young working-class Brits ascribe to deviant and risk behaviours on holiday in one particular resort. Using the life testimonies of five young men from Southside, and numerous focus groups, interviews, informal conversations and observations, I have tried to deconstruct different elements of this type of tourism and offer an alternative understanding which considers popular culture, *habitus*, larger political economic forces and subjectivity in the constructions of deviance and risk. I have gone beyond conventional understandings of surveys (Bellis et al., 2000; 2003; Hughes et al., 2004; 2009) and dated constructions of subculture because a more nuanced account – one which considers the micro-subjective against the macro-structuro and macro-culturo – is necessary to understand this issue in its entirety. The main point I have tried to stress is that their deviant and risk behaviours are not the fault of some pathology or defunct psychological attribute but more about how this cohort have come to self-validate behaviours which are expected of them over time and, as a consequence, how these are subtly guided and endorsed by a social context which is only interested in making money at their expense.

A reflexive methodology

I realise that some of the methods I have used in this book may be considered to have been ethically unsavoury in that when I was researching intoxication, I was also involved, to some extent, in the drinking games and associations of the Southside Crew and others in the holiday scenery. I also witnessed drug use, drug dealing, and violence. It is my deep belief, however, that there are no universal guidelines on how a researcher should behave in these situations where his or her participants are engaged in regular, heavy alcohol consumption and drug use. How a researcher deals with the challenges posed by an extreme research environment can be justifiably argued on the basis that they need to 'do what they need to do' to get the data without 'doing too much of it' (if you know what I mean). I feel the knowledge produced in a study such as this, as long as it is intended to fall within the scope of academic work, means that the validity of its conclusions should satisfy the demand of critical assessment.

If only I could convince some of the world's leading drug and alcohol journals which have outright rejected my work because they considered it to be 'unscientific' or have accused me of 'enjoying myself' (see Young, 2011). I wonder sometimes if academics understand that these kind of ethical problems come intricately wedded with the research territory of ethnographic research with young people who are in hedonistic celebration. I only say this because throughout the fieldwork, and even the writing-up process, I continued to receive tiresome jibes from some colleagues about this being 'a holiday' or 'dream research trip'. I was seeking to understand a problematic social issue and, in doing so, if I am honest, cannot say I took much 'fun' from it. The ugly truth is that to go in cold sober into somewhere like the San Antonio West End is not 'fun' at all because you feel invisible; you feel like a nothing who can't appreciate this everything. To be immersed in this life was like a constant 'below life' vision for me (Chapter 10) because seeing how this feature of contemporary life operates above and below the façade of this glittery ideology was not pretty; especially given that millions of people make this kind of holiday exodus each year and a significant number of whom return with problems as a result.

A conclusion: Deconstructing deviance and risk abroad

In the wake of uncertain life trajectories and faltering class identity, appears for this cohort of young working-class Brits the safety net of consumption, leisure, play and the 'good life', complemented by a premise of making impulsive decisions to spend money/experience 'life'/do 'crazy' things. At home, these people are used to drinking, taking drugs and playful forms of deviant and risk-taking as part of their leisure time and so therefore it should be no surprise that these practices are often what they initiate when they can find any time out of the home predicaments in spaces which are specifically and symbolically designed for their blowout. Because the holiday is a finite period of time, it must be seized and must be milked as much as possible before the inevitable return to the banality of routine; the fixed time period often acting as a rationale to power through, especially when a self-deconstruction starts to pull apart some of the perceived failings of home life which only amplifies the need to make the most of it. Abroad, they are anonymous and a new permissiveness is personally rationalised and socially buttressed which allows for experimentation and exploration of the deeper realms of their fetishes and fantasies in the resort space which happily matches these requisites. It doesn't matter that things go wrong because 'what happens' are seen as stories for life and this time won't come around again. Therefore while the behaviours, and even some of the crude consequences, are constructed as something *real* which they can tell back home, they are as much a by-product of an ideological social conditioning of *being* over a period of time as they are the result of the way this cohort are drawn into excess, deviance and risk-taking by the global corporations, commercial entrepreneurs and tourist companies/organisations.

My book therefore shows there is no pathology at play in the behaviours that this group of British tourists exhibit abroad as they are in a perpetual cycle not only to consume and live by this consumption as a means of identity construction, which by the way constrains them, but are equally bound to transgress it as a way of attempting to abscond its hold on their life. Here play the roles of the weekend, festivals, mega-events, and, in the context of my work the holiday: but this is the tight grip the social system has on its subjects. Deviance and risk behaviours abroad are therefore as:

- much culturally embedded as they are reinforcing;
- much socially expected as they are situationally engaged, and consequently become as;
- much subjectively reasoned as they are structurally and spatially encouraged.

Thematic and theoretical discussions

My book attempts to take forward areas of analysis in the context of British working-class young people, the elasticity of 'youth', tourism, political economy and identity. There now follows a more detailed analysis of these thematic and theoretical points.

The pleasure class

Numerous commentators have attempted to conceptualise the changing nature of contemporary working-class populations across Western countries. The movement from modern to postmodern society has dislocated this group from work, family and community (Giddens, 1991), instead replacing it with rampant individualism, and commitment to leisure and consumption practices (Bauman, 2007) to fill the subjective ontological void. In Guy Standing's work, many of the people in my book would fall under the banner of 'The dangerous class or The Precariat'; a sort of lumpen population of people who live by extremely uncertain work and labour futures and are increasingly economically, socially and structurally excluded by the neoliberal order. Slavoj Žižek, on the other hand, may put some of this same group into the category of the 'consumtariat' as they are a population who have been drained of their political consciousness and class identity, and instead seek self-actualisation through participation in consumption practices and the symbolism of social envy which is consequently generated (Hall et al., 2008). I agree with these hypotheses.

However, I feel these theoretical concepts usefully incorporate agentic leisure pursuits which derive from individual decisions made by this group under the collective and ideological guise of neoliberal consumer capitalism. This, as I have shown, is intrinsic to how they see themselves and what they expect from themselves (Chapters 7, 8 and 9).

I would like to introduce a more useful term to explain this working-class group. I think the people in my book represent a *pleasure class*; one which displays the hallmarks of being politically pacified (Hall et al., 2008), structurally uncertain (Standing, 2011) and bound to consumption as a means of *being* (Žižek, 2011). This is because, given that these elements have made them more collectively impotent and individually powerless, as a response, leisure and the adoption of the imagery of popular culture becomes increasingly more important as a life pursuit (Chapter 5), which also reflects the movement towards a consumer society (Chapter 3) – the ideological safety net to help divert attention from the impending problems of postmodern life. While Veblen's (1994) notion of the *leisure class* posited that conspicuous consumption assisted in the elite's status maintenance through access to economic capital and the capacity to spend it willy-nilly, the *pleasure class*, which occupy the lower realms of the class structure, are equally trapped in this perpetual labyrinth of consumption; seeking to exhibit a greater class superiority through excessive spending, except these days, the pressure to attain status by these means is rampant (Hall et al., 2008). Similarly, like the *leisure class*, my cohort display an *unfreedom* (Žižek, 2002) as they are locked in an eternal cycle of sustaining their status through market participation (Veblen, 1994). To do this, many spend money they don't have, amass significant debt, and borrow from people just to participate in opportunistic moments motivated by immediate gratification and play (Rojek, 2005) – all of which are ideologically constructed as what they should be doing to be living fulfilling lives (Chapter 5).

As we have seen, a way of seizing control in the material world and its restrictive operations is to transgress this *unfreedom* by means of *hyper-consumption*; an exaggerated form of consumption which is engaged to reassert individual ontology and retrieve some sense of 'freedom' (Chapter 7). Yet despite the potential ugly consequences (Chapter 9), it is seen as fun and so is the life by which others around them live and the cultural imagery of the celebrities who do the same (Chapter 5). Indeed, the cultural ideology of this life doesn't seem to be about work but 'play', about making the most of it considering the prospects (Young, 1999). Life for many people in this bracket becomes about seeking *pleasure* (sensation) through *leisure* (means by which sensations are constructed). Most of the people in this book 'live for the weekend', and in general, by any means which will offer them an opportunity to escape the normative boredom which they construct as the everyday experience (Presdee, 2000). Similarly, on holiday they are under exaggerated social conditions to feel as if they need to reproduce a level of spending to participate in social life; to feel culturally included. When I offer this term *pleisure class*, I'm not undermining the structural or cultural conditions which play a part in this group's social position and *habitus* but suggesting that this is how they respond; by trying to look good, getting good deals on things and creating memories – by seeking pleasure.

The holiday career

In his psychological work in tourism studies, Pearce (2005) refers to the 'travel career' – how primary motivations to travel evolve over time as people get older and how new tastes develop. I feel that this term feels much more of a middle-class reference of experiencing the world and certainly doesn't really feel relevant to the people in my sample who don't 'travel' per se; they are not 'travelling'. This term seems to suggest that one is on the move, exploring and experiencing (O'Reilly, 2000) and this isn't really what they are doing because they are bound not only by their *habitus* to reproduce behaviours abroad but also by the spaces of the resort. The people in my book 'holiday'; they are holidaying because they generally don't go anywhere of cultural significance outside the resort and have little ambition to learn the local language/culture. Nor do they really build any other taste for anything other than getting hammered unless they manage to claim 'holiday capital' (Chapter 7).

From what I can see therefore, this book is the first piece of evidence to suggest that these young people embark on a series of experiences of excessive consumption abroad over the life project which often result in deviance and risk behaviours (*The long road* in Chapter 6) in different resorts which offer the same sort of thing. From a young age, they are strategically guided by the cultural ideology of popular culture about the holiday: what it is all about, what one should do and where one should go (Chapter 5). When they first embark on a group holiday in their late teens, it is normally a Club 18-30 package where they are soldiered into excess by tour reps (Hesse et al., 2008); they are socialised and coerced into learning what is expected of them in the resorts and this early exposure to the good life often results in deviance and risktaking (Helen in Chapter 9). Tourist companies design these holidays around concepts of 'youth' and 'sex' and market them as 'something to be done' in life (Pritchard and Morgan, 2005). What they experience on these holidays aligns perfectly with their *habitus*, which has been shaped by similar processes, and often provides the impetus for these people to explore other resorts which are reputedly 'more classy' or that have 'better music' but often only cater for their maturation rather than any new class position. In fact, there seems to be no obvious class upgrade as many end up exhibiting the same sort of behaviours but in different places; it's just they are a bit older and might have a bit more money to spend.

Ibiza is one such pinnacle of the holiday career but in recent years, largely because of image problems and increasing competition with other new European resorts, tourist numbers have reduced (IREFREA, 2010). This book shows that new forms of commodification and commercialisation have been initiated to capitalise on tourist spending in the short time they spend on the island. This, I feel, is aptly visible in the aggressive marketisation of the island over the last decade reflected in the TV programmes, radio discourses, advertising campaigns, and music chart entries. Ibiza, its music, the DJs, the Superclubs, etc. have flooded media channels and magazine covers which have quickly found a way into the perceptions of impressionable young people (Chapter 5) who are supposedly 'free' to choose what life has to offer and make the most of 'new experiences' and get the life-tick list completed. So, like those who have taken the *long road*, there are increasingly a group of *short cutters* who now take their virgin holiday in Ibiza (Chapter 6). There they learn that excessive consumption offers and satisfies a temporary means of 'feeling free' as well as an ideological social status to accumulate if they return.

My research shows just this; that this *long road* is now more of a *short cut*. And the more experienced crowd who took that long road don't like the younger, inexperienced group potentially infecting the space so start to distance themselves in other tourist spaces; by elevating themselves to the bigger, better Superclubs or beach hotels; or may even look to work in Ibiza for a season, 'living the dream' (Bourdieu, 1984, Chapter 6). However, importantly, it doesn't matter too much how they have come to arrive because they will come back because some say it is 'the best' even without knowing what is good about it; they discover there are more things to do which with their gain come increased ideological social status, and there – Ibiza and its tourist spaces – becomes reconstructed as a place of nostalgia (Chapter 10). With more Ibiza experience comes the need to attend more select events like opening and closing parties with perhaps a holiday between. They feel they have arrived; that nothing can crown this experience because there is so much more now to see and do the next time they come back. But the reality is they haven't found the rainbow's source and are unlikely to because the global corporations, commercial and marketing entrepreneurs and tourist companies consistently introduce new exclusivities to spaces to accommodate these ambitions for 'new experiences' and 'higher status' (Chapter 8).

While a few of these people have tried to elevate themselves to higher cultural grounds like 'travelling' having reached their mid-20s, it doesn't seem to match with their *habitus* which has been shaped around what they do on weekends (Chapter 5) as well as carefully constructed ideologies of what the holiday should be all about (Chapters 1 and 6). There are adventures outside the resort which they consider to be 'cultural' but, more often than not, these attempts reinforce what they are *not* rather than what they are trying to *be*; they *are* what they *are* so they better get back on the lash quick before it starts to get out of their comfort zone. And once the Promised Land has been experienced, it is difficult to leave which is why as some mature, and even have families, it becomes difficult to take a step down to another resort. Not to worry, there is always a potential week-long stag party to be had there or something for which the rationale to return can be justified (Chapter 10).

'Youth', the holiday and time folding

These days, the 'holiday' is not only something which occurs once a year but is now embedded in our everyday practices, interactions and social exchanges. Equally, the holiday is not only just about time out but now about excessive consumption. Furthermore, there are not only whole industries behind it but a whole set of interactions and discourses which, under the premise of the current cultural framework, support the idea that 'one needs to go' to places like Ibiza before they die because everyone is talking about them. In the context of Ibiza, this is the ideology of the Superclubs, the branding, the style and status which allegedly comes with it. However, what this work shows is that, unlike some other popular resorts for Brits, Ibiza is as much constructed around the concept of 'youth' as it is 'nostalgia.' For young Brits as well as those maturing, the reason to go is often guided by global corporations, marketing entrepreneurs, the music industry and further influenced by the spatial commodification processes which leaves endlessly indelible the notion that a particular social status can be attained through careless spending power (Meethan, 2001, Chapter 8). These discourses find their way into the individual psyche of these British tourists who often end up thinking:

1. 'I am young, and this is what young people must do to enjoy their youth so we better do it before we get older and get responsibilities'.

Or:
2. 'I am a bit older [or maybe not] but have all these responsibilities in my life and so deserve time out and need to make the most of the short period of time I have away from a pressured life'.

Consequently Ibiza becomes a space where *time folds* and does so in two ways: (1) it is where generations converge to do the same sort of thing, only at different points in their lives; (2) and also becomes the personal rationale to return as people seek to rekindle memories – as a consequence of the *commodification of nostalgia*. My book shows also that the holiday career, as a process of maturation through package holidays and marketed destinations, is built around finding another spot where the 'younger crowd' or the 'chavs' don't go: a place to distinguish their age and more supposedly refined attitudes (Bourdieu, 1984, Chapter 6; Hayward and Yar, 2006). As this cohort mature, they seem to feel out of kilter with the younger generation so it makes sense to go to a place where age demarcations are less clearly defined – such as Ibiza – where there are 'others like you' who are 'your age' or thereabouts who are doing similar things to you. It becomes permissible and possible, even if you did it many years ago. And consequently, it sets in action the rationale to reproduce similar behaviours in the name of nostalgia. What further assists in the personal impetus of returning to Ibiza is the way in which it is commercially constructed as a 'place of dreams' as much as a place to 'rekindle those dreams' reflected in the CD compilations, Ibiza Superclub nights out in London and reunion parties (Chapter 10). So a *commodification of nostalgia* assists in the *time-folding* process (Chapter 6) as the *short cutters* look to recall last summer's madness as much as the seasoned Ibiza goers who took the *long road* or the middle-aged stag party seek to roll back the years. After all, the West End has all manner of bars, pubs, clubs to cater for all ages and musical interests.

Group dynamics on holiday: Gender, class and transgression

The book has presented data which shows how holiday group dynamics function. I have shown how for the *pleasure class* it is normal to get drunk and high (Parker et al., 2002; Blackman, 2011; Hadfield and Measham, 2009) as it is to engage in risk behaviours which result from this consumption (Measham, 2006; Wilson, 2006; Measham, 2008). This is because the weekend has, for most, become a popular bracket where ritualised practices of drinking and drug-taking occur (Blackshaw, 2003) as a consequence of the constructed 'banality of the working week' reinforced by radio, TV and other cultural mediums (Chapter 5). It's therefore not surprising that, in the absence of work identities (Rojek, 2005), opportunism and seizing the moment (Standing, 2011) become the order of the day as this group say they live only for the weekend, engaging in excessive consumption (Presdee, 2000; Hayward, 2004) as a means of self-realisation (Schor, 1992) to generate some social kudos (Standing, 2011; Hall et al., 2008; MacDonald and Shildrick, 2007) – even if it is painful (Lacan, 2006; Hall et al., 2008). The weekend therefore becomes a time when gendered and class relations are played out (Rojek, 2005) and this is often reinforced by similar popular constructions of nights out at home and on holiday evident in TV dramas, films, and public disclosures of celebrity lifestyles (Blackman, 2011). Intoxication, image, sex, and violence are therefore part of the cultural repertoire in the context of the holiday which the *pleisure class* absorb (Chapter 5) and this is evident in the way groups of friends hype up the holiday through Facebook, pub conversations and prepare for it by shopping and going to the gym. There is a pressure to look nice, to portray an image and 'youthfulness' so others will be envious of their bodies or the expensive clothes they fashion (Bunton et al., 2011; Hall et al., 2008) – just as Pamela said they need to look like they belong in Ibiza.

My work shows that these aspects of *habitus* come to the foreground when these groups holiday abroad – one which resides around *hyperconsumption* and exploratory behaviour into the leagues of the extreme. The *pleisure class* are tourists because they playfully celebrate, and unlike travellers seek a comfortable time away where things are done/organised for them; they seek the old and familiar (Skey, 2011; O'Reilly, 2000). However, because of this, the time they have to maximise their experiences is reduced which means what will be done will likely be of greater importance – an occasion to be seized (Meethan, 2001). It is these dynamics which tend to sustain a pressure of *hyperconsumption* abroad, most evidently presented in the testimonies of the Southside Crew throughout the book. On holiday, time and space are compressed and this assists in the pressure to *hyperconsume* and engage in the bizarre and extreme behaviours in the group context in an effort to create memories and stories which can be reflexively accessed to help brighten the home tedium (Griffin et al., 2009, Chapters 9 and 12). But the *pleisure class* have come to know that this social scenery is full of people like them in groups similar to theirs having watched things like *Club Reps, Ibiza Uncovered, The Villa and Sun, Sex and Suspicious Parents* or *Kevin and Perry Go Large* or *The Inbetweeners* (Chapter 5) or having previously had similar holidays (Chapter 6). All this supports the expectation that what they do there is inherently normal (Chapter 7).

In the excitement of this 'liberation' from home routine (Chapter 7), groups like the Southside Crew reinforce the need to party. Indeed, among both gender groups, individuals are frowned on for not keeping up or in line with what is expected but male groups tend to exhibit more competitive elements of group excess. My work shows that among both gender groups, passive members have little choice other than to keep up with the pace of consumption – they should not be the outsider to the fun or they'll be a party pooper. And while the more bizarre and extreme deviant and risk-taking activities generate *jouissance* – a pleasure from the pain of the experience (Lacan, 2006) – it is both male (the Surreal Brummies in Chapter 9) and female groups (the girls beaten by police and bouncers in Chapter 9) who display this commitment.

The holiday resort also offers the means for men to performatively sustain masculine identities (Winlow, 2001), and while many of the men in this book try to carry off a 'hard man' exterior, having invested heavily in gym time or even steroids (such as Paulie), they also display their capacity as men to keep up with *hyperconsumption*. They are thus executing their masculine embodiment as part of a wider construction of identity and the gendered self (Thurnell-Read, 2011). The other side of this embodiment is the way men see women as commodities – as objects to be consumed (Andrews, 2005; Rojek, 2005; Briggs, 2012). This is as much the female tourists (Sinead in Chapter 9) as it is the lap-dancers, strippers and prostitutes (Princess and Precious in Chapter 8). But this is hardly surprising when the ideology of sex as a marketing strategy is continuously beset upon male groups like the Southside Crew which reinforces their quest to 'shag a bird' (Chapters 8 and 10). In San Antonio, women are seen as trophies to be won; just like on *The Bachelor* and *Take Me Out* (Chapter 5). In fact, in some of the testimonies of male groups, there seems to be a competitive edge to sleeping with as many 'birds' as possible – in one group, five young men had placed bets before the holiday to see who could have sex with the most women. In this way, gender relations are ideologically reproduced abroad (Andrews, 2005, Chapter 5).

And while men are encouraged to deploy some sort of hard exterior masculinity, as the nights progress, hegemonic masculine positions are suspended (Bakhtin, 1984) and instead leisure practices start to influence masculine identities as men's relations become playful, exploratory, homo-erotic and almost self-mocking as deviance and risk-taking becomes the order (Thurnell-Read, 2011). Recall the testicle licking and guilt-free, full-on kissing of the Surreal Brummies. By comparison, the female groups don't seem to display this attribute; instead some tend to offer more protectiveness over each other, a few often feeling guilty in hindsight of deviance and risk (Thomas, 2005), while the male groups almost encourage it to feed group bravado (Thurnell-Read, 2011).

However, the research also shows that both gender groups are at risk of separation over the course of the night out – consciously perhaps this manifested in arguments as holiday tensions surface in the self-deconstruction process (Chapter 7) but perhaps subconsciously through a narcissistic individualism (Hall et al., 2008). This is because what lies in the shadows of group collectivity – no matter how many years they go back as friends – is an individual subjective intention of self-reward, self-indulgence and the need to create envy (Chapter 3) and I have seen first-hand how these agendas and subjective adventures into *jouissance* interfere with group solidarity. Recall how George got jealous that some of his mates had girls in his room which led to a rift in holiday relations (Chapter 10). It is these kind of fissures which can result in dispersal from the group on the West End on a night out. In the case of the Southside Crew, they separate because there is some indirect disagreement to buy the Clubland package (Chapter 9). For some time, Jay and Nathan tag along with the bar crawl but it isn't long before their motivation and interest dwindles, and, even shortly after, they separate and make their own way back to the hotel – as do Marky and Paulie. Nathan, in particular, ends up defecating while he ejaculates as he receives oral sex from two prostitutes. For Sinead, separation from the group came from petty arguments and when there was a conflict of music style (Chapter 10). The fact that these groups have such diverse interests is, in my view, a consequence of the mass commercialisation of Ibiza and how it draws less homogenised groups to the island that have within them different interests, conflicting holiday agendas and subjective desires. And it is this separation in spaces like the West End that puts them at increased vulnerability of other risks specific to the social context such as getting completely lost, self-injury, having unprotected sex, being robbed, or even worse, death (Chapters 8 and 9).

Space, class and consumption: The spatial commodification of status stratification

Holiday spaces have historically had class dimensions (Chapter 6) where conspicuous consumption takes place (Veblen, 1994). These days, those very same dimensions exist in Ibiza; that is, tourist spaces operate as containers for human interaction and consumption (Miles,

2010) and are commercially commodified in an attempt to divide the classes. I have shown how this operates for the majority of my sample on the West End which is principally and symbolically designed for practices in line with their *habitus*, while to some extent, a degree of distinction (Bourdieu, 1984) is constructed in the other resort of Platja d'en Bossa where prices are a little higher. The Superclubs, Eden and Es Paradis (situated in San Antonio) act as the fallback for the *pleisure class* when they don't have the capital to do places like Pacha and Amnesia. If they do make it to those clubs, they may even feel out of place just as they might if they make 'cultural trips' outside the resort – it can quickly confirm who they are *not* while at the same time remind them who they shouldn't try to *be*. Even within the Superclubs there are defined spaces according to available capital such as the VIP lounges. The private hotel of Ibiza Rocks (in San Antonio) is not quite Ushuaia (in Platja d'en Bossa) but it offers the *pleisure class* the same thing: a contained space where everything is on their doorstep and with their attendance comes some degree of social status above people who have a cheap hotel on San Antonio Bay. The same goes for the beach clubs such as Ocean Beach Club Ibiza (San Antonio) and Blue Marlin (near Platja d'en Bossa) that determine which class of tourist can attend by spending power, but also within that, offer that same tourist a scale of social status from which to buy into. This is also reflected in the resort space of San Antonio which acts as symbolic reference points which bolster an imagined sense of solidarity and belonging (Billig, 1995; Skey, 2011) so, for some like Liam, being elsewhere feels awkward.

Many people in this book seek to elevate themselves up these invisible status ladders by attending the bigger clubs which have attached to them higher prestige (as well as higher prices). In this respect, even when the hallowed ground is experienced, there are further levels to navigate evident in the numerous levels of VIP within the spaces. I have tried to show how the *spatial commodification of status stratification*, on one hand, allows the social classes a reassuring detachment (Bourdieu, 1984), while on the other, creates an intra-class strain within that same class bracket (Hall et al., 2008). I mean that a 'status inadequacy' can set in when people realise there are ideologically constructed higher gradations attached to social status and thus more steps to climb on the invisible ladder. This is because the ideological impetus has been embedded in these tourists, regardless of class, that spending equates to the 'experience'. However, the commercial side of this is that there have been strategic moves to maximise spending as tourist numbers decrease (Payeras et al., 2011) because of resort competition (Chapter 1), and it is this which has led to the commodification of space which has attached to it new levels of social status. It gives people a glimpse of what they can be with further participation and money but really it is a rainbow-chasing process; a journey to a destination in which no one arrives.

And because the same spaces contain the tourists and their spending (Chapter 8), draining other local tourism businesses in the process (Chapter 4), the battle for tourist spending becomes volatile. It's difficult to see how this takes place between the Superclubs because featured DJ nights overlap so hence there is no competition to recruit numbers and because of the volume of international tourists. However, this is perhaps most notably visible on the West End where numerous PRs physically direct British tourists into bars for tempting alcohol deals while local Spanish business owners can only scratch their heads and shrug their shoulders; they may as well join the bandwagon and open up a bar or café serving fried breakfasts, and thus banal nationalism is reinforced (Billig, 1995).

Because these class dimensions are visibly defined in these spaces, it is difficult to see how concepts of subculture and 'neo-tribe' are useful for my cohort. Firstly, there seems to be no subcultural resistance in what is happening in Ibiza but rather mass conformity; a universal culture which displays consumer culture lifestyles (Chaney, 2004; Miles, 2000). Ibiza these days is not about hippies or the 'Balearic beat' enthusiasts – it is about everyone. It is about the masses. It is about getting one over on the 'other' by saying they have 'done Ibiza'. The attraction is not only the music but the club, the aesthetic design of the club, the dancers, the entertainment and the style which is connected to the clubs but more importantly the 'names' and the branding (Meethan, 2001; Miles, 2010; Ritzer, 2000). Equally, universal expressions of togetherness of the 'neo-tribe' in the context of club crowds (Maffesoli, 1996; Malbon, 1999) which have been said to dissolve structural divisions such as class (Redhead, 1993) don't seem useful to conceptualise the way in which class inequality determines access to certain spaces via spending power. Just look at how each Superclub has its own VIP area but also the way in which beach hotels and beach clubs organise a seemingly unlimited variety of VIP (Chapter 8).

Capitalismo extremo and the ideology of 'living the dream'

What I think we are witnessing in the resort of San Antonio and across Ibiza is an extreme form of capitalism – *capitalismo extremo* (Chapter 9); a sublime money-making process led by the global corporations, commercial entrepreneurs, tourist companies/organisations who ideologically pied-piper these tourists into excess, deviance and risk – all at the expense of themselves. It is this influence which directs spending, leisure and lifestyle choices (Ritzer, 2000) and thereby contributes to the global machinery of consumption (Bramham and Wagg, 2011), seeking only to advance established political and economic interests (Rojek, 2005). Such is the power of *capitalismo extremo* that it convinces individuals that excessive consumption equates to *being* in the world even if it means risking their health and wellbeing. This keeps them happy, and keeps them participating in the system – this is their reward. Under this framework, people think they are 'free' but in participating, they only reinforce how *unfree* they are (Žižek, 2002); they are conforming to what is expected of them, what they have come to learn and therefore sustain the very ideology which subconsciously blinkers them to what is going on. They say if it weren't for the pillars of routine and responsibility, that the holiday is true 'freedom' and allows them to cast themselves adrift to enjoy the splendours of the commercial market yet the psychological moulding of *capitalismo extremo* is clever in that it not only makes them hungry for this participation but thirsty for the moment in which to do it. So while they seek to break free from constraints of consumer society on holiday by making *real* something from its *unreal* hold over their ontology (Žižek, 2008; Žižek, 2011) they only end up conforming to it through *hyperconsumption* and by taking behaviour to the limits of the bizarre and extreme.

And business comes when people are in the mood for intoxication; through the formal economy (taxis, fairground rides, club entry) and the informal economy (prostitution, drugs). Recall how Paulie was drunk, did a bungee jump while taking an E and had it filmed – all for the experience, all for social kudos on YouTube and Facebook. So there is a blatant capitalisation on the personal need for self-indulgence and need to feel special as well as the group moment to celebrate and create memories (Griffin et al., 2009). In Ibiza, spending equates to some sort of status elevation. Yet memories come at a price (if they can be remembered at all). This is a social system set up to tell people they are special when they go out clubbing, get their nails done and that the road to happiness is found through impulsive spending acts and considering only 'number one'; reflected in the MasterCard mantra of 'priceless memories' and Chris Brown's club hit 'Beautiful people' in which the lyrics blaze out *'live your life, live your life'*. These elements help to secure individualistic attitudes of 'seizing the moment' and haphazard spending as a means to a fulfilling life (Bramham and Wagg, 2011). Ideology coats norms and value systems of the *pleisure class* to think they should go to Ibiza and do these things yet some can't quite identify why they necessarily went (Chapter 7).

Of course not all British tourists go to Ibiza and go crazy but most, if not all, the people in this book don't see past the glittery façade to this stylistic consumption which has a darker side to it. What do I mean by this? Robberies, violence, hospitalisations, self-injury and in some cases death. This is the side of Ibiza people don't learn about nor does it want to be known. During the course of the research, there were numerous accounts of people dying on holiday; I passed the small flower shrine of one in Chapter 8 while in another account during my time there a dead Brit was found floating naked in a hotel pool up the road, no passport, nothing. Holiday gone very wrong. Another Brit beaten and raped, woke up on a beach with only her ripped bikini to cover her as she found her way to her hotel, which she wasn't sure where it was because it was her first night. Not one for Facebook, really. No one cares about these people and the party goes on. Remember Mark from Chapter 9? He had been in Ibiza for three weeks, most of which he had slept rough. He spent £2,000 in Ushuaia within a few days of his arrival, lost his accommodation, got beaten up three times and had half his ear bitten off by a bouncer. He has no money to return to the UK and didn't know where the consulate was. Yet when I bought him breakfast, he smiled at me saying he was *'living the dream'*.

And consider the casual workers whose work is play and whose play is work (Guerrier and Adib, 2003). They seek the glitz and glamour and set off looking for work in Ibiza with a puss-in-boots attitude to be a DJ or a club dancer, but many end up in debt and some get into awkward situations where they have to get involved in drug dealing or start dancing in strip clubs where they are likely to be manhandled by aggressive staff if they don't play by the customers' demands. They engage in large amounts of excess and risk-taking (Hughes et al., 2009; Briggs et al., 2011b) because their work is play and their play is work and this seems to make their lives only more precarious and uncertain – consider Tim's journey in the last chapter (Standing, 2011).

And it doesn't stop there because the campaign for tourist spending is everywhere! What about the trafficked women from Nigeria and sub-Saharan African countries who line the streets each night in San Antonio looking for a little business? A few euros for a blow job? If you are feeling like bargaining, they will go all the way for €20 or anal sex for €30. What about the abuse, violence and police harassment they are subject to? And the fact they have to survive like this having escaped political persecution in their home countries as well as navigate structural and social exclusion not forgetting the continual stigma in the playground of the resort? This is the everyday in Ibiza. This is the real Ibiza. People don't know about it because their time there is finite and no one is interested in what goes on behind the flashing lights as it is time for fun and self-indulgence.

Many people in my book cannot recognise how *capitalismo extremo* lures them in, chews them up and spits them out. And now the allure to Ibiza has caught the attention of a younger, more impressionable crowd who have little understanding of what is required of them. Like the Inexperienced lads and Unpredictables in Chapter 5 and Lizzy and her friends in Chapter 9, they spend more, go home penniless, wonder why it was worth it but still express the desire to come back next year for the 'good life' or to 'live the dream'. One can't visit Ibiza without acknowledging the palpable vulgarity of commodified and painful excess; millions of young Britons enjoying themselves so much it hurts (Lacan, 2006; Hall et al., 2008). No one in the leisure and tourism industries is interested in making sure people do things safely; these industries operate at the expense of draining these young tourists of every single penny they have while away on holiday.

The internal tug of war on identity

The holiday is a unique moment in space and time in which the people in my sample rationalise that certain practices and behaviour become possible. They say they are 'free' from the constraints of routine and responsibility in the adult playground of San Antonio and engage in a carnivalesque celebration of 'time out of time' (Bakhtin, 1984; Redmon, 2003) and, like Nathan, indulge in vulgar bodily pleasures, take-aways, drinking and sexual promiscuity (Featherstone, 1991). The spatial and situational transition made from home to holiday produces in most an immediate shift in identity as the world of the resort is perceived to be anonymous and a *new permissiveness* is constructed around behaviour. These people are in a space which is not reflective of what they do back home and where social relations are more random (even though they are with their friends). The new territory (resort) and social moment (holiday) means new experiences can be gleaned, like Nathan trying laughing gas on the West End (Chapter 9) – even air is commodified! (Harvey, 2005) It is this attitude which often leads to experimentation with particular drinks/drugs, the self and social exploration of bizarre fetishes and playful deviance (the Surreal Brummies in Chapter 9) as well as extreme risk-taking (Tim and the Kent youth in Chapter 10) in holding spaces which are strategically designed to accommodate behavioural fetishes (Miles, 2010) as well as turning a blind eye to the consequences because they prop up the economy (Chapter 8).

However, as people withdraw from their home routines and responsibilities, the 'weekend self' finds life on a daily basis (Tutenges, 2012). There then begins for many a process of self-deconstruction and reflection as the self unknots itself from life back home. For most, what follows is more of an awkward confrontation with home life rather than an escape because it triggers a life evaluation (Beck and Beck-Gernsheim, 2002) of where they are, what they are doing, who they are with and what they will do. When these achievements are juxtaposed for validation against a hegemonic cultural framework of 'be your own boss' (Standing, 2011) and someone who has the material showpieces of a fulfilling life (Hall et al., 2008) – good job, nice car, house, etc. – they realise they fall short, often feeling inadequate and powerless at the same time. They don't quite know what to do because they can't initiate their dreams (Chapter 7) like the people on *Britain's Got Talent* can (Chapter 5). This personal stymie prompts most to rationalise that they need to seize what they can while they are in Ibiza and this aids in the process of *hyperconsumption* (Chapter 7) and consequently deviant and risk behaviours (Chapter 9).

In addition, this transition from overcast home life to the excitement of unlimited hedonism further prizes open this reflexive channel, in which the euphoria of what is before them collides with who they are back home: this starts to skew home realities which have now been warped by the false life of the resort. This identity reversal reinforces the perceived 'shitness' of 'normal' life but also, at the same time, exaggerates the false happiness of the 'good life'; the *hyperconsumption* providing a bogus safety net, an imagined escapism towards security and this is easily stimulated by the need for subjective ventures into *jouissance* (Lacan, 2006, Chapter 3). As self-deconstruction gathers momentum, the ideology of the holiday life becomes ever more appealing. Some start to convince themselves that Ibiza really is 'the dream' so why can't they achieve those dreams at home like the people on *X Factor*? (Chapter 5) A few may make self-promises to do certain things to alleviate their position when they return but the allure of the 'good life' in Ibiza starts to win them over and an internal psychological tug of war starts to ensue made possible through a *reflexive double door* which nakedly reveals to them the steep-hilled tasks of improving their home circumstances; they see at the same time their past as well as gaze at their future. Most, like Liam and Graham in Chapter 10 feel powerless to instigate change. The *reflexive double door* therefore swings both ways and reveals the life left behind, as the life which lies ahead; it presents the overcastness of then and the colour of here; the impossibility of before and the possibilities of now. And it is here where the ideology temptingly starts to sell to them the idea of returning to Ibiza or staying for a permanent holiday to 'live the dream'.

This tension reaches its climax as the holiday starts to draw to a close and many experience a 'below life' vista. For many, there is an internal confusion over what is the real reality as a personal ontological abyss starts to present itself as the potential realisation dawns on them that what they have just participated in could have been a nothingness which was pointless. There now seems to be a little doubt about who they are. Why is this? Well, because Ibiza is all about ideology, material life, and has become to be known as something which is *said* to have been done which indicates that its value is for group approval in social discourse – to have been like one of the elite or celebrities or like on the TV programmes or films (Chapter 5) – rather than inherent ontological value. And because it exists in this material sense through social discourse to generate kudos/create envy like this, it makes *'below life'* feel even deeper, even heavier on the psyche. This is precisely why some people leave in the psychological doldrums, feeling hollow. In an effort to cushion a deep fall into this personal abyss, many quickly rekindle their commitments to the excessive consumption on weekends on return and even rebook flights out to Ibiza.

Some recommendations

Getting to grips with this type of tourism, and reducing the negative impact it has on the local Spanish community and the tourists, will require significant structural, social and cultural change. It is unlikely to happen in the near future but if my book has managed to persuade policymakers, tourist companies or even tourists themselves to think differently about how holidays are experienced in places like San Antonio, then my job has been done.

Rebalancing Ibiza

This book has shown how one particular resort area, and the tourist demographic surrounding it, has significantly shifted over the last 30 years. But it is not only this shift which has affected the tourist demography in Ibiza for, with this tourism, come significant burdens to the infrastructure of the island. This is exacerbated by, what is seen to be, a limp criminal justice system in its almost permissive response to drugs because without providing a more coherent and strategic policing agenda to deal with these issues, the system provides an open invitation to deal in drugs – and it is so easy. Indeed, even when it is clear that some parts of the local authority have tried to engage with elements of the NTE (Superclubs in particular) in an effort to reduce drug-related problems, progress has been hampered somewhat. The outlook also looks bleak in light of what this book presents in instances where new avenues for endorsing deviant and risk behaviours have evolved in the wake of previous concern about them. However, I feel it was only necessary to at least expose this initially before we start to think of solutions.

I don't think we can ignore the power of neoliberal consumer capitalism in how San Antonio functions as a resort. I say that in the context of free market politics which permits the permeation of profiteering at the expense of the tourist. When I say 'expense' I mean how their financial, practical and personal health can be damaged in these resorts. Couple the holiday attitude towards a blowout and some degree of experimentation and playful deviance with clever marketing schemes and commodified nothings for sale, and you see people lose all their money and become homeless, some turning to drug dealing just to get them home; you see people injure themselves because they drank too much but it is equally the decision-making of the individual as it is the coercion of the group and the environmental invitations to get wasted. This is what is happening in Ibiza and I suspect most resorts across the Mediterranean: it is private organisations and tourist companies who are laughing all the way to the bank while the local authorities have next to no control over the way in which these places operate yet still sanction their business.

Change will rely on curbing global corporations and British proprietors from taking the bulk of business and reconsider local equivalents which can help sustain the economy. However, as foreign investment and business start to predominate and tourist numbers from other countries start to dwindle, the need to sustain the San Antonio economy has entered a new phase; the quest to retain business has become desperate. To adjust, the resort regenerates (Chapter 4) in an effort to retain the remaining hegemonic tourist body – the British. So as the cancer has set in and started to eat away at all other viable economies, it has left San Antonio helplessly reliant on the very forms of tourism which causes problems. This is why, in many ways, Ibiza has entered what seems to be an irreversible phase of its political economy in that, on one hand, although it claims distance from the behaviours which supposedly tarnish its image, it has become, at the same time, almost subservient to those behaviours because they derive from the haphazard spending of tourists such as the British. This incestuous relationship is evident in all sectors of the economy: tourism, law enforcement, and health. Well, perhaps we shouldn't be surprised because the gates to business in Ibiza have been thrust open to free market enterprise and reversing this tide relies on undoing already-established deals with global corporations, marketing companies and tourist organisations as these are the very industries and services which now overtly advertise Ibiza and therefore bring capital to the island.

Yet Ibiza has moved into new tourist territory: numbers are decreasing while, at the same time, there is pressure to maintain the economy in the six months it is open for business against an increasingly free enterprise in which British companies/organisations come in and cream off profit from local residents and businesses. If the politicians and people of significant power on the island are remotely interested in redressing this imbalance, they will also look to find new tourism ventures during the winter months so the economy does not rely wholeheartedly on what is made during the summer. This will mean such an aggressive marketisation needn't take place and, as a consequence, tourist numbers will balance out over the year rather than in six months. They will also need to reconsider how law enforcement responds to resort areas and relatively allocate resources accordingly, while at the same time, reconsider the strategic responses of the different levels of the police.

Reducing the risks of the tourists

At the moment, aside from the frontline charity (Ibiza 24/7) which works in San Antonio, there seems to be no real attempt on the part of the tourist companies, the British government and social institutions to provide realistic interventions to reduce the risks associated with drinking/taking too much drugs/engaging in general excess. Even the local authorities and, to some extent without fault law enforcement, look pretty limp in their response. So where do we start? Probably the easiest way of redirecting practices away from such excessive forms is to start to regulate the tourist companies who profiteer from the excess they endorse to the young people who embark on their holidays. This would do two things: (a) in the long run, reduce the familiarity of deviance and risk on holiday and, as a consequence (b) hopefully act as some means to redirect their attitudes away from reasoning that these sorts of holidays confirm what they should be doing when they are abroad. If the tourist companies selling these holidays were more responsible, especially with the way in which their staff pied-piper people into excess, deviance and risk, but at the same time, fail to be responsible when things go wrong, then some problems could be avoided. Just like the FCO, the emphasis is that people are responsible for themselves, which of course is true, but when those very same people are at the mercy of commercial forces which are capitalising on their attitudes to spending, then the contours of 'responsibility' become blurred. In this respect, the FCO, the British government and consulate will need to regulate commercial enterprises which take venture in Ibiza while, at the same time, work in tandem with the local authority to reduce environmental risks associated with the San Antonio resort and the Superclubs.

This research has shown that we are talking about companies and organisations which are clearly taking advantage of the attitudes towards excessive consumption which the British tourists bring on arrival in places like Ibiza. Similarly, the airline companies who are also tempting people into the holiday spirit by offering alcohol deals and blatant advertisements to start the party on the plane will need to be diminish this marketing. The research points to the risk of the first night so at every stage – airport, plane and in the resort – endorsements to party should be reduced which may reduce intoxication levels. Notwithstanding, the level at which intoxication is sustained is often high risk throughout the holiday. People power through their excess for days on end. A greater harm reduction presence, such as that as Ibiza 24/7, could help reduce problems if they had more resources to deliver messages about over-intoxication and the risks of particular drugs.

More stringent policies need to make accountable the Superclubs to provide free water to British tourists (as well as others) so that people don't dehydrate in the clubs. Similarly, some sort of monitoring needs to take place in the new private spaces for consumption such as the booze cruises and beach hotels (Ibiza Rocks, Ushuaia) and beach clubs (Blue Marlin and Ocean Beach Club) which have evolved as a consequence of the policies introduced to clamp down on afterparties. In these spaces, the risk continues to be prevalent evident from this book. In the resort context, the local authorities and police need to work closer together to regulate the British (and other foreign) businesses which seem to operate so freely and attract numerous PRs and even British tourists looking for work. When they come, many don't find work and end up dealing drugs as a means to get by. The PRs, as we have seen, are central to the function of the drug market and often deal direct to tourists on booze cruises, in bars and clubs in and around the resort. If the authorities were more proactive in regulating these companies, fewer PRs/tourists would have the opportunity to deal drugs, although it is acknowledged that the demand for drugs may not necessarily diminish.

Perhaps some of these recommendations are unrealistic but I would hope that anyone who reads this book would at least understand that pathologising exercises on the deviant and risk behaviours of this group of British tourists are equally unhelpful and that, in fact, there are wider dynamics at play which influence and direct the same cohort into those practices. The chances are high that this form of tourism will continue for the foreseeable future as we continue down the pathway of neoliberalism, guided by the promise of more profit and wealth at the expense of a widening subsection of society who live by increasingly precarious means. To give you an idea of how things are continuing in the wake of what I have written, the final chapter charts some of the pathways of the Southside Crew since the summer of 2011.

12
Meanwhile across the Mediterranean ... (or So Some Wish)

Introduction

As I conclude this book in December 2012, I have known the Southside Crew for about 18 months. In that time, Paulie split up with his girlfriend five weeks after returning from Ibiza in August 2011; he continues to try to avoid drug dealing when he can as the police attention has increased in Southside. He also went out to Ibiza again in the summer of 2012. Marky retreated from the group, claiming he hadn't got much money to come out with them on Southside. Over the New Year into 2012, Jay and Nathan got into a dispute over £80 which Marky didn't pay back to them. He is still in his relationship although much of Southside seem to think his girlfriend is cheating on him. Jay turned his hand to amateur boxing and had a few fights to 'prove himself and man up'. Like Nathan, he still works on a temporary basis in the construction industry but has dipped into drug dealing on a few occasions to earn extra money. Fortunately for both Nathan and Jay, they received suspended prison sentences for their violent misdemeanours outside the club and didn't go to jail in the end. Nathan became a father but split from his girlfriend. He also failed to pay his court fees (on top of the £6,000 debt he had) and, at the time of writing, was also wanted by the police for a drunk and disorderly charge (as they also found out he gave a false name when arrested). Although a depression and hard drug use has started to set in, he is undertaking a psychology course and shows more enthusiasm by the day to try to take his life forward.

Shattered dreams of 'Magaluf 2012'

When I call Nathan one evening in May 2012, he is on his way to the gym. He puts me on speakerphone and in the background are Jay and

Paulie. They start to get a bit lairy and start cajoling over our time in Ibiza. Nathan, who has been giving a false name and job situation at the gym to get a discount, asks when next I am coming down and we agree another night out soon. They also tell me that Simon has gone out to Ibiza for a week but had only booked a hotel for one night; Jay had to pay for a hotel with his credit card from the UK when Simon phoned pleading for some help. How could he not have known? We recount some stories from Ibiza, and between the laughs and jeers, it gets serious when they confess that, despite some months of planning, they will unlikely go to Magaluf. A few days later, and in a series of text messages, it seems their holiday plans have been scuppered:

Dan 9.37 p.m.: Nathan, Hope u keeping well and little one is OK. Want to meet up? Btw, u in Majorca or Ibiza this summer?

Nathan 9.40 p.m.: Simon's gone to Ibiza today. Jay is not allowed to go [because he was unfaithful to his wife] and I'm skint so probably won't go there unless I wheel and deal [grow cannabis and sell it] … Love to get out there. I get my [driving] licence back in a couple of weeks so cud do it after u get back.

Dan 9.44 p.m.: This is my last trip! What happened to your licence? How long is Simon out for? Is he working?

Nathan 9.48 p.m.: Lost it four months ago because I crashed my truck. Ouch! Simon is out there until Friday, meeting loads of people out there like professional footballers. I think he is just following the crowd like a sheep though ha! He was working for Jay's dad on a house job. We may go to Tenerife, all of us and the missus. Love to get back out there.

Simon's return to Ibiza

As the summer of 2012 begins, Simon leaves for Ibiza where it seems he has taken behaviour to another level:

Jay: But Simon has been off his fucking face on ketamine but he doesn't normally do it but all the other lads are into it, like the lads he is with out there, they do 14 or 15 pills in a day and ketamine and he told me recently, because he is out there now, he did some ket and woke up six hours later with his

	face on the couch and his muscles were in some spasm, he was like 'ahhhhh' and his muscles had cramped and locked.
Nathan:	And he got fucked in the arse.
Dan:	Right.
Jay:	They did some fucked-up shit to him. A mate put his dick up his [Simon's] arse.
Dan:	I kind of want to know more, but I don't. Am I hearing this correctly?
Nathan:	Then they shaved someone's pubes off.
Dan:	Were they fucked up?
Jay:	Completely. Simon was telling us that one guy was standing on the edge of the balcony going 'aaaaahhhh' because he was so high and one couldn't go out the last night because he was so paranoid and couldn't go out. They left him on the roof as he couldn't sleep and when they came back the next day he was still shivering.
Dan:	Why is he doing this now?
Jay:	He is showing off, he is trying to prove himself.
Nathan:	Yeah, they was, well one guy was fucking Simon up the arse and said it was a bit of banter, weren't it.
Jay:	Like these are normal guys, not poncey. Like last year when they went away, they were shitting in water bottles, shaking it up like diarrhoea and throwing it at each other. They do shit like that.
Dan:	That makes you look like gentlemen!

And on one summer night in Southside ...

It has been about a month since I saw members of the Southside Crew, but the day has finally come. I am to meet in a nearby town, go back to Nathan's place to get ready for the night out, meet the rest of the Crew and head out on Southside. Nathan tells me I have to *'look nice'* because we are going to a *'sophisticated place'*. I feel under pressure because when I open my wardrobe, all I see are the same things I have been wearing for years. A mixture of fashion laziness and changing style modes makes me feel very outdated and old as I pick out a fake Versace shirt, some smart grey trousers which are a little tighter than I remember and some new black shoes. At least I look as if I am actually going out on the town but this time I can't hide the glinting grey hairs at the sides of my temple or ignore the wide M shape which represents my receding hairline.

As I drive down to Southside, memories of Ibiza and our time start to resurface again. I meet Nathan and we drive to his flat, where he quickly disappears into the shower to get ready, leaving me to watch a large flat-screen TV while the picture jumps from time to time. When he comes out of the shower in a towel, he walks in on me removing my trousers and shirt to change for the night ahead. He then starts showing me some scars on his arms. Thinking they are connected to his life in the army, I say they must have hurt and this is when he tells me that he did them to himself. A few months ago, after splitting up with his ex-girlfriend Pam and getting lonely in the flat, he bought razors and self-harmed then bought 60 paracetamol and tried to commit suicide. He ambles back into the bathroom and beckons me in, asking if he should shave around his goatie. I borrow some of his David Beckham aftershave and try to think of a way to make the pathetic excuse for the hair on my head look reasonable. We call Jay, agree to meet in a nearby village and head off in the car.

We drive into a village just outside of Southside where he grew up and he points out where his divorced parents live. We park up next to Tesco and, as we walk into a bistro pub, there sits what looks like Jay … but is his brother Dave. Nathan tells me as Dave goes to the fruit machine that he is pretty much welfare dependent and *'well on the weed.'* As Dave somehow wins £70 on the gambling machine only to then reinvest it and lose it, I sit over a few pints and catch up with Jay and Nathan. We then amble into Tesco's and meet up with Mike, a forty-something year old man, divorced with three kids. His shift finishes at 10 p.m. and they consider waiting around for him but Nathan tells them to hurry up. Dave and I then argue quite loudly about whether I should drink or not; he keeps insisting we should get a taxi and saying things like 'Come and stay on my sofa mother fucka, let's get smashed'. Meanwhile, Jay and Nathan purchase a bottle of Disaronno and coke as their conversations echo around Tesco's.

As we walk out, I go to withdraw cash. Looking over my shoulder, the antics have begun and they are cajoling by my car. Jay pours half the coke on the pavement and then adds the whole bottle of Disaronno to the half-filled coke bottle. As we get in the car, Dave requests a stop-off at his place to collect a spliff. Nathan turns on Radio 1, turns it up loud and winds down his window. As we take a detour into a run-down area in another town, they argue about which is the quickest way and I do a few circuits of the roundabout waiting for them to decide on our direction. They start to moan that Marky and Paulie aren't coming out. Apparently, they were like 'Yeah we're coming out tonight' but backed out at the last minute; a common trait of late they say. Jay then tries to persuade me to drink a few more beers but I will surely be over the limit. When I say no, he recalls the first time he was arrested for drink driving when he was 17.

When we arrive at Dave's, we all wait in the car for a few minutes for him to collect a large spliff. He returns smoking it and its aroma quickly fills my car. Jay covers over the silence by offering the Disaronno cocktail around. We then get onto the subject of Ibiza and they all cheer and recall how it was a good night. 'Consider we didn't know each other then, mate, it was a good night. It was like we was best mates' says Jay. 'He's got shitload of qualifications, mate' says Nathan to Dave. The music continues to bang out as we raise our voices to talk over the noise from the cars as we have our windows down; the sun setting slowly on the horizon. They finish the bottle of Disaronno between them and Jay changes his tune; the reminiscing clearly buoying the night out ahead.

When we finally arrive in Southside, after what seem to be a never-ending maze of roundabouts and life stories, we pull up in a car park on the seaside. The breeze is welcome as we all get out of the car and our full bladders force us to find some improvised location to urinate; Dave and Nathan piss on a van next to my car while Jay and I choose a more discreet alleyway. Nathan almost runs ahead as if to anticipate meeting his girlfriend but really it is a strategy. As Dave catches up with him, Jay tells me this is a better idea. Because they are notorious for problems in Southside, especially disorder and violence, Nathan will go into the club with Dave and Jay and I then will follow shortly after. As they wander in, the bouncer gets cautious of Dave as he sort of wobbles all over the place under the heavy influence of cannabis and beer. Fortunately Nathan shepherds him in and the bouncers instead get distracted as some women in high heels come to the door.

As we walk up the stairs, fashionable, smart and stylish people of a much broader age group come down, laughing. The club is some sort of wannabee exotic zone with palm trees and sunset-type lighting; it could easily be a club in Ibiza. The room is set around a gigantic, oblong bar in the middle where staff mostly seem to be serving cocktails and shorts. I decline an alcoholic drink but Jay insists on a shot of sambuca so I neck it. Dave tells me I should be getting wrecked and insists I should drink more and stay at his with Jay and Nathan. 'Champagne will change your mind ... let me get the champagne!' but my lack of interest seems to kill his suggestion as he takes another gulp of his beer while Nathan and Jay follow up with a double Disaronno and coke.

They seem now to be in quite flirtatious moods, especially Nathan who is expecting the imminent arrival of his girlfriend. As in Ibiza, Nathan seems lured by numerous women who pass and Jay also calls out to them as they pass while Dave just stands there smiling, swaying and taking large swigs of his beer. Yet unlike Ibiza, sexually charged comments are absent and their efforts to flirt are less direct, perhaps because of the 'sophisticated scenery' and the familiarity of Southside. While Nathan starts talking to a woman who has tattoos on her neck, Jay tries to get me in the mood for dancing and swings his chunky arm around my neck. We move in time to the music but I am not quite sure of the dance moves he is trying to coerce me into and it ends up looking like badly timed groin thrusting moves augmented with finger points to the ceiling. They all then buy more shots. When I get back from the bathroom it is Nathan who then disappears. Jay and Dave then launch a small attack on Nathan, joking how he used to give blow jobs to other boys and that he is gay (yet as we dance shortly after, Jay is the one who is slapping my bum and cuddling and kissing me on the neck).

When Nathan returns, I recall how I still have some photos from Ibiza of us drinking in a bar and show him. As I flick through them, Jay cheers as his grip on my shoulder turns into a potential headlock. After a few more dances, including one by Rihanna titled 'We found love in a hopeless place', there is an awkward scenario as a group of middle-aged women start dancing near us. In the drunken chaos that is their dancing effort, one who has been eyeing Nathan drops her drink only for it to spill halfway down the back of a man behind Nathan. Unfortunately, this man is an absolute beefcake (tall and well built). As the man turns around, Nathan picks up the glass at the same time and the beefcake thinks it is Nathan who has been careless with his drinking and they start exchange verbals which quickly turns into a face-off. Nathan descends into a very serious cold zone as their chests start to meet and I come between them, guiding Nathan away as if we were about to start ballroom dancing together.

It is well past 11 p.m. when Mike – the transformed, well groomed and nice-smelling Tesco worker – bounces into our dancing space. Immediately he starts dancing with Jay and gyrating against the ladies who just pass while they generally continue uninterested. As Dave seems to take more of a side-of-the-dance-floor role, Mike drifts off to talk to him. By now, further shots have been purchased and Jay's energy is starting to grow. He has somehow created a space of a few square metres around him as he randomly swings his arms around and side-steps his feet in something which he calls dancing. He smiles at me but I wonder how long it will be before he either accidentally knocks someone out or accidentally knocks a drink out of someone's hand, gets in a fight and then knocks them out.

After Nathan and Jay get further drinks, Nathan tries to slip his fingers up a girl's skirt and blames it on a drunken-looking Dave when she turns around. Nathan gets away with it and just in time as well it seems because seconds later, his girlfriend Anna arrives. This is the moment that it gets a bit odd between Nathan and Jay because as soon as she arrives, Jay and Dave leave the area. Apparently, Jay gets upset if the lads are out and they have to accommodate a 'bird'. For the next 20 minutes, Nathan just kisses and politely fondles his girlfriend (nothing from diaphragm upwards or bellybutton downwards). Mike and Jay beckon me outside for a cigarette. As I go out onto the club balcony, which overlooks the beach and pier, the chill of the night air hits me along with cigarette smoke and loud laughing. The club's balcony is jam-packed with fashionable-looking people smoking, giggling and kissing. Nathan and Anna then follow. She and her friends, all with kind of mousy personalities, stand around and Nathan comes over with his mobile phone in his hand and starts showing me texts from another woman, Pam, his ex-girlfriend.

When we all go inside, Nathan breaks from Anna as she starts talking to her friends and comes over to me: 'I've got to get out of Southside, mate' he whispers as he puts his arm around me. While others jump around I start looking at the series of text exchanges: Pam is coming to pick him up from Southside and she is the one who he *'really feels for'*. He drags me over to the club exit, where Jay, Dave and Mike are lingering. It seems that they are up for another venue but know not of Nathan's dilemma. Nathan disappears to the toilet to call Pam while Mike and Dave go outside. I am then left with quite a drunk Jay who says he wants to wait for his 'boy' [Nathan]. Jay and I descend down the stairs arm in arm; his short-sleeved shirt still flapping open as he whistles at the female groups that enter. As we wait on the stairs, Jay rests his sweaty head on my shoulder and I smile politely at the people who come up and down the stairs, perhaps thinking that our night is already at its conclusion given Jay's state. He comes alive again when Nathan bounds down the stairs. It seems he hasn't told Anna he is leaving yet she is still in the club.

As we all linger outside the club in the sea breeze, Nathan makes another call to Pam which prompts Jay to start shouting in the phone '*fucking bitch*' – so he obviously knows the story too. Pam left Nathan to be with her ex from South Africa but they recently split up a week ago. Because Nathan was hurt months ago by Pam, Jay got defensive and told him to '*man up and move on*' telling him she was a '*fucking bitch*'. Jay pulls my arm and says 'Come on Dan, there's a club there, let's go' and in the background Dave and Mike wave me over. I say I have to stay with Nathan and he gives me a bear hug and kisses my neck but shouts aggressively '*chatting shit mate*' at Nathan who continues to talk to Pam on the phone.

As we get into the car, Nathan's phone starts to ring and ring and ring. It stops for a few seconds ... then rings and rings and rings. As we leave Southside, it finally dawns on Nathan what he is doing when Anna, who is the one who has been trying to call, sends him a text: 'Babe, what are you doing to me on my birthday?' it says. The consequences of his actions start to sink in and to avert the silence between us and distract his guilty conscience he puts on the radio. The lively dance music doesn't seem to help; 'What am I fucking doing?' he says to himself, yet the guilt seems to be forgotten for short periods as he excitedly directs me to another town where Pam lives. He thinks up a story to account for his sudden disappearance from the club which involves him getting in a fight with the bouncers and then being subsequently arrested.

When we arrive at Pam's, Nathan invites me in saying she won't mind. There is some sort of party going on, and when we go in through the door, I am greeted by a man called Steve. Walking into the open-plan apartment, the place is a tip. Dust, food and rubbish litter the floor and a mixture of empty beer cans, spirit bottles and take-away boxes stack the coffee table, chairs and most surfaces. The kitchen area is practically unrecognisable and is full of more empty alcohol cans and bottles and dirty cutlery. Around our feet tiptoes a little grey kitten which Pam later tells us cost £100. Pam and her friend Char are in the toilet where Char is throwing up; we all wait in the living area trying not to look at the shithole around us but find no other conversation. When they stumble out, Nathan lurches towards Pam in her revealing short dress and starts to try and kiss her. I sit on the sofa with Steve while he smokes a cigarette. Char then drags her chunky body through the rubbish, kicking a KFC bucket out of the way, and sits down in a pile with a glass of wine. I go to the toilet which is filled with vomit and a used tampon. When I return, Pam questions Nathan on his sudden escape from Southside and wonders why he couldn't wait for her to arrive. I intervene and say that I had to go back to London so thought it was best I take him and this seems to help. 'Are we still goin' out?' says a very drunken Char as she sits with her head in one hand and a cigarette in another. Well, they are all dressed for it. I can see where the evening is going for Nathan as he starts to cuddle and kiss Pam and make my excuses to leave. Nathan follows me outside. We embrace and he says *'love you mate'*. It is around 2 a.m. The night is still young and so are they.

Notes

1 Introduction

1. Public Relations (PR) workers come under the broad umbrella of young Brits working abroad as 'casual workers'.
2. There are inevitable discussions vis-à-vis distinctions between 'tourist' or 'traveller' and class interpretations of the 'holiday' of which I challenge in Chapter 6.
3. For a similar analysis of the spatial construction of gender relations in the context of British holiday-makers see Andrews (2009).

2 The Flexible but Entirely Serious Methodology

1. Ellis and Bochner (2000: 742) define autoethnography as 'autobiographies that self-consciously explore the interplay of the introspective, personally engaged self with cultural descriptions mediated through language, history, and ethnographic explanation'.
2. I speak Spanish and some of these interviews were conducted thus.

4 Ibiza: The Research Context

1. A musical style.

5 Goin' Ibiza: Home Lives and the Holiday Hype

1. 'Brits abroad', 'British holidaymakers', 'British tourists abroad' were three main keyword searches entered into *The Guardian* (n = 1928) and *The Sun* (n = 683) online newspaper websites.
2. See Adams (2000).
3. See *The Guardian* (2000).
4. See Green (2001); see also Da Costa (2001) for an article written the very same day in the very same paper which promotes the main Ibiza clubs.
5. See Gillan and O'Hara (2001).
6. See Tate (2012).
7. See Pooran (2012).
8. N-Dubz was a rap/pop group which has attained some success in the UK charts in recent years.
9. See Tinney (2012).
10. Verbatim quotes taken from the very first episode.
11. A reality TV series in which competitors have to 'survive' various different circumstances, often having to complete tasks to gain commodities or things which will assist their survival.

12. The comments used here are selected as some 'status entries' and 'wall posts' generated a significant number of responses.

7 You Can Be Who You Want to Be, Do What You Want to Do: Identity and Unfreedom

1. At the time of writing, Queens Park Rangers were a Premier League football team.
2. Interestingly, it reminds me of when I went to Morocco when I was 18 on a package holiday and the tour reps said 'Don't go outside of the resort, the locals can be a bit starey [can stare at you]'. I bet the Moroccan locals were saying the same about the busloads of white tourists: 'Don't go in the resort, they can be a bit starey'. However, it was said in such a way to scare the tourists while at the same time ensure that they would take trips out with their operators.

8 The Political Economy: Consumerism and the Commodification of Everything

1. Gary Lineker, a famous former professional footballer and TV sports presenter in the UK.
2. See *The Sun* (2012).
3. In one blog interview, one club rep says that 'It's better to have no sleep and be massively stressed out than be in Manchester, in an office when it's raining'. The good life trumps the redundant life with ease. Despite the pain and stress, it's all 'worthwhile'.
4. See Nicholas (2008).
5. In Spain, the *Policía Local* have limited powers and are found in every town/city of more than 5,000 people. Generally, they are responsible for traffic, parking, monitoring public demonstrations, guarding municipal buildings, and enforcing local ordinances. Generally policing rural areas and highways, the *Guardia Civil* are responsible for responding to numerous social issues which include counter drug operations, customs, illegal smuggling, security and terrorism and intelligence gathering. The *Policía Nacional* mainly patrol urban areas and deal with criminal justice issues, terrorism and immigration.
6. There are other such hotels and beach clubs but for this study, only these were visited.

Glossary

If you go out in UK town centres and/or go on holiday to places like San Antonio in Ibiza, you will likely come across these terms.

All-nighter: Drinking/taking drugs all night into the early hours of the morning

Bender: Can mean a homosexual but in this research is mostly referred to a lengthy drinking/drug-taking session

Bird: Young woman

Blow job: Oral sex

Bollocks: Refers to something which is 'unbelievable' or can mean men's testicles

Booze: Alcohol

Brazilian: A waxing to remove pubic hair around the bikini line

Buzzing: A good feeling/experience from a substance

Chav: Derogatory term used to refer to young people in the UK who lack class/style/taste

Chillout: To relax or take it easy; also referred to zones such as 'chillout place'

Cock: Penis

Coke: Cocaine

Cunt: Used as term of endearment as it is an offensive word. It is also a derogatory term which refers to the vagina

E: Ecstasy tablet

Fag: Cigarette

Fella: Person

Fuck it/Fucks you up/Fucked/fucked-up: Why not do something/negatively affects someone/to have had sex with someone or be heavily intoxicated/to be unfair, make a mistake or be heavily intoxicated

Geezer: Person

(The) 'Good life': All the culturally prescribed rewards in life

Grand: One thousand pounds (£1,000) or to mean feeling OK (e.g. I'm grand = I'm fine)

Hammered: Getting drunk, but can also mean generally intoxicated

Innit: isn't it? (question tag)

Ket: Ketamine

'Living the dream': The pursuit of the 'good life'

LOL: Laugh out loud, commonly referred in text messages and social media

MDMA: Ecstasy

Messy: Getting drunk, but can also mean generally intoxicated

Minger: Derogatory term meaning 'ugly girl/woman'

Minging: Getting drunk, but can also mean generally intoxicated

Missus: Wife/girlfriend/partner

OMG: Oh, my God!

On it/get on it: Start the process of becoming heavily intoxicated through alcohol and/or drugs

Pills: MDMA/Ecstasy

Pubes: Pubic hairs

Ponce/poncey: Derogatory term used to describe homosexual men/to act as a homosexual man

Pussy: Derogatory term which refers to the vagina

Riffraff: 'Scum' or to be of low social calibre

(To be) ripped: Have a nice body

Ripped off: To lose out financially to something/someone

R&B: A music genre that combines elements of rhythm and blues, pop, soul, funk and hip-hop

Session: A lengthy period of time dedicated to intoxication

Shag/Shagging: To have sex/having sex

Skinny dipping: Swimming/paddling naked in the sea

Skint: To be low on/without money

Slag: Derogatory term used to describe a loose or easy woman

Spliff: Form of roll-up cigarette which may include drugs like cannabis

STI: Sexually Transmitted Infection

Superclubs: Large, multistorey, high-capacity, high-profile nightclubs. In this study, they are Pacha, Amnesia, Space, Privilege, Eden and Es Paradis.

Tits: Breasts

Wanker: Can mean idiot or someone who masturbates

Wankered: Drunk

Weed: Marijuana

References

Andrews, H. (2005) 'Feeling at home: Embodying Britishness in a Spanish Charter resort', *Tourist Studies*, Vol. 5 (3): 247–266.
Andrews, H. (2009) '"Tits out for the boys and no backchat": Gendered space on holiday', *Space and Culture*, Vol. 12 (2): 166–182.
Andrews, H., Roberts, L. and Selwyn, T. (2007) 'Hospitality and eroticism', *The International Journal of Culture, Tourism and Hospitality Research*, Vol. 1 (3): 247–262.
Bakhtin, M. (1984) *Problems of Dostoevsky's Poetics*. Trans. Emerson, C. London/Minneapolis: University of Minnesota Press.
Bataille, G. (1967) *La part maudite*. Paris: Les editions de minuit.
Baudrillard, J. (1998) *The Consumer Society: Myths and Structures*. London: Sage.
Bauer, R., Kormer, C. and Sector, M. (2005) 'Scope and patterns of tourist injuries in the European Union', *International Journal of Injury Control and Safety Promotion*, Vol. 12: 57–61.
Bauman, Z. (2007) *Liquid Times: Living in an Age of Uncertainty*. Cambridge: Polity Press.
Beck, U. (1992) *Risk Society: Towards a New Modernity*. London: Sage.
Beck, U. and Beck-Gernsheim, E. (2002) *Individualisation*. London: Sage.
Bellis, M.A., Hughes, K., Bennett, A. and Thomson, R. (2003) 'The role of an international nightlife resort in the proliferation of recreational drugs', *Addiction*, Vol. 98: 1713–1721.
Bellis, M.A., Hale, G., Bennett, A., Chaudry, M. and Kilfoyle, M. (2000) 'Ibiza uncovered: Changes in substance use and sexual behaviour amongst young people visiting an international night-life resort', *International Journal of Drug Policy*, Vol. 11: 235–244.
Billig, M. (1995) *Banal Nationalism*. London: Sage.
Blackman, S. (2007) '"Hidden ethnography": Crossing emotional borders in qualitative accounts of young people's lives', *Sociology*, Vol. 41: 699–716.
Blackman, S. (2011) 'Rituals of intoxication: Young people, drugs, risk and leisure' in P. Bramham and S. Wagg (eds) *The New Politics of Leisure and Pleasure* (pp. 97–119), London: Palgrave Macmillan.
Blackshaw, T. (2003) *Leisure Life: Myth Masculinity and Modernity*. London: Routledge.
Boorstin, D. (1992) *The Image: A Guide to Pseudo-events in America*. New York: Vintage Books.
Bourdieu, P. (1984) *Distinction: A Social Critique of the Judgement of Taste*. Cambridge, MA: Harvard University Press.
Bramham, P. and Wagg, S. (2011) 'Unforbidden fruit: From leisure to pleasure' in P. Bramham and S. Wagg (eds) *The New Politics of Leisure and Pleasure* (pp. 1–11), London: Palgrave Macmillan.
Briggs, D. (2010) 'Researching dangerous populations: Some methodological reflections', *Safer Communities*, Vol. 9 (3): 49–59.

Briggs, D. (2012) *Crack Cocaine Users: High Society and Low Life in South London*. London: Routledge.

Briggs, D. and Turner, T. (2011) 'Risk, transgression and substance use: An ethnography of young British tourists in Ibiza', *Studies of Transition States and Societies*, Vol. 3 (2): 14–25.

Briggs, D., Turner, T., David, K. and De Courcey, T. (2011a) 'British youth abroad: Some observations on the social context of binge drinking in Ibiza', *Drugs and Alcohol Today*, Vol. 11 (1): 26–35.

Briggs, D., Tutenges, S., Armitage, R. and Panchev, D. (2011b) 'Sexy substances and the substance of sex: Findings from an ethnographic study in Ibiza, Spain', *Drugs and Alcohol Today*, Vol. 11 (4): 173–188.

Bunton, R., Crawshaw, P. and Green, E. (2011) 'Risk, gender and youthful bodies' in W. Mitchell, R. Bunton and E. Green (eds) *Young People, Risk, and Leisure: Constructing Identities in Everyday Life* (pp. 161–180), London: Palgrave MacMillan.

Busher, J. (2012) "There are none sicker than the EDL": Narratives of racialisation and resentment from Whitehall and Eltham, London' in D. Briggs (ed.) *The English Riots of 2011: A Summer of Discontent* (pp. 241–261), Hampshire: Waterside Press.

Calafat, A., Blay, N., Bellis, M., Hughes, K., Kokevi, A., Mendes, F., Cibin, B., Lazarov, P., Bajcarova, L., Boyiadjis, G., Duch, M., Juan, M., Magalháes, C., Mendes, R., Pavlakis, A., Siamou, I., Stamos, A. and Tripodi, S. (2010) *Tourism, Nightlife and Violence: A Cross Cultural Analysis and Preventive Recommendations*. Palma de Majorca: IREFREA.

Carlson, R., Wang, J., Siegal, H., Falck, R. and Guo, J. (1994) 'An ethnographic approach to targeted sampling: Problems and solutions in AIDS prevention research among injection drug and crack-cocaine users', *Human Organisation*, Vol. 53: 279–286.

Chaney, D. (2004) 'Fragmented culture and subcultures' in A. Bennett and K. Kahn-Harris (eds) *After Subculture: Critical Studies in Contemporary Youth Culture* (pp. 36–48), New York: Palgrave MacMillan.

Coffey, A. (1999) *The Ethnographic Self: Fieldwork and the Representation of Identity*. London: Sage.

Corrigan, P. (2010) *The Sociology of Consumption*. London: Sage.

Crawshaw, P. (2011) 'The logic of practice in the risky community: The potential of the work of Pierre Bourdieu for theorising young men's risk-taking' in W. Mitchell, R. Bunton and E. Green (eds) *Young People, Risk, and Leisure: Constructing Identities in Everyday Life* (pp. 161–180), London: Palgrave MacMillan.

DHS (German Centre for Addiction Issues) (2008) *Binge Drinking and Europe*. Hamm, Germany: Achenbach Druck.

Ellis, C. and Bochner, P. (1999) 'Bringing emotion and personal narrative into medical social science', *Health*, Vol. 3: 229–237.

Ellis, C. and Bochner, A. P. (2000) 'Autoethnography, personal narrative, reflexivity: Researcher as subject' in N. K. Denzin and Y. S. Lincoln (eds) *Handbook of Qualitative Research*, 2nd ed. (pp. 733–768), London: Sage.

Featherstone, M. (ed.) (1990) *Global Culture: Nationalism, Globalisation, and Modernity*. London: Sage.

Featherstone, M. (1991) *Consumer Culture and Postmodernism*. London: Sage.

Ferrell, J., Hayward, K. and Young, J. (2008) *Cultural Criminology: An Invitation*. London: Sage.

Geertz, C. (1973) *The Interpretation of Culture*. New York: Basic Books.

Giddens, A. (1991) *Modernity and Self-identity: Self and Society in the Late Modern Age*. California: Stanford University Press.

Govern de les Illes Balears (2011) *Sistema estatal d'informatició permanent sobre addició a drogues. Resultats de l'indicador d'admissions a tractament per consum de substàncies psicoactives a les Illes Balears*. Majorca: Govern de les Illes Balears.

Griffin, C., Bengry-Howell, A., Hackley, C., Mistral, W. and Szmigin, I. (2009) '"Every time I do it, I absolutely annihilate myself": Loss of (self-)consciousness and loss of memory in young people's drinking narratives', *Sociology*, Vol. 43: 457–476.

Guerrier, Y. and Adib, A. (2003) 'Work at leisure and leisure at work: A study of the emotional labour of tour reps', *Human Relations*, Vol. 56 (11): 1399–1417.

Hadfield, P. and Measham, F. (2009) 'England and Wales' in Hadfield, P. (ed.) *Nightlife and Crime: Social Order and Governance in International Perspective*. New York: Oxford University Press.

Hall, S. and Winlow, S. (2005a) 'Anti-nirvana: Crime, culture and instrumentalism in the age of insecurity', *Crime, Media, Culture*, Vol. 1 (1): 31–48.

Hall, S. and Winlow, S. (2005b) 'Night-time leisure and violence in the breakdown of the pseudo-pacification process', *Probation Journal*, Vol. 52 (4): 376–389.

Hall., S., Winlow, S. and Ancrum, C. (2008) *Criminal Identities and Consumer Culture: Crime, Exclusion and the New Culture of Narcissism*. Cullompton, UK: Willan.

Hallsworth, S. (2005) *Street Crime*. Cullompton, UK: Willan.

Harvey, D. (2005) *A Brief History of Neoliberalism*. Oxford: Oxford University Press.

Hayward, K. (2004) *City Limits: Crime, Consumer Culture and the Urban Experience*. London: Glasshouse.

Hayward, K. and Hobbs, D. (2007) 'Beyond the binge in "booze Britain": Market-led liminalization and the spectacle of binge drinking', *The British Journal of Sociology*, Vol. 58 (3): 437–456.

Hayward, K. and Yar, M. (2006) 'The "chav" phenomenon: Consumption, media and the construction of a new underclass', *Crime, Media, Culture*, Vol. 2: 9–28.

Hawkes, S., Hart, G., Shergold, C. and Johnson, A. (1997) 'Risk behaviour and STD acquisition in genitourinary clinic attenders who have travelled', *Genitourin Medicine*, Vol. 71: 351–354.

Hesse, M., Tutenges, S., Schliewe, S. and Reinholdt, T. (2008) 'Party package travels impact on alcohol use and related problems in a holiday resort – A mixed methods study', *BMC Public Health*, Vol. 8, No. 351.

Hobbs, D., Hadfield, P., Lister, S. and Winlow, S. (2003) *Bouncers: Violence and Governance in the Night-time Economy*. Oxford: Oxford University Press.

Hollands, R. (2002) 'Divisions in the dark: Youth cultures, transitions and segmented consumption spaces in the night-time economy', *Journal of Youth Studies*, Vol. 5, (2): 153–171.

Hughes, K., Bellis, M.A. and Chaudry, M. (2004) 'Elevated substance use in casual labour at international nightlife resorts: A case control study', *International Journal of Drug Policy*, Vol. 15: 211–213.

Hughes, K., Bellis, M.A., Whelan, G., Calafat, A., Juan, M. and Blay, N. (2009) 'Alcohol, drugs, sex and violence: Health risks and consequences in young British holidaymakers to the Balearics', *Addiction*, Vol. 21 (4): 265–278.

Hyde, A., Howlett, E., Brady, D. and Drennan, J. (2005) 'The focus group method: Insights from focus group interviews on sexual health with adolescents', *Social Science and Medicine*, Vol. 61: 2588–2599.
IREFREA (2010) *What Can We Do to Prevent Violence and Harms among Young Tourists? Facts and Practical Recommendations for Healthier and Safer Holidays*. Palma de Majorca: IREFREA.
Katz, J. (1988) *Seductions of Crime: Moral and Sensual Attractions of Doing Evil*. New York: Basic Books.
Kelly, D. (2011) 'The sexual behaviour and sexual health needs of young British casual workers in an international nightlife resort'. Presentation at Club Health conference in Prague, December.
Kuntsche, E., Rehm, J. and Gmel, G. (2004) 'Characteristics of binge drinkers in Europe', *Social Science and Medicine*, Vol. 59: 113–127.
Lacan, J. (2006) *Ecrits*. London: Norton.
Lea, J. (2002) *Crime and Modernity: Continuities in Left Realist Criminology*. London: Sage.
Lloyd, A. (2013) *Labour Markets and Identity on the Post-industrial Assembly Line*. Ashgate.
MacDonald, R. and Shildrick, T. (2007) 'Street corner society: Leisure careers, youth (Sub)culture and social exclusion', *Leisure Studies*, Vol. 26 (3): 339–355.
Maffesoli, M. (1996) *The Time of the Tribes: The Decline of Individualism in Mass Society*. London: Sage.
Malbon, B. (1999) *Dancing, Ecstasy and Vitality*. London: Routledge.
Measham, F. (2004) 'The decline of ecstasy, the rise in "binge" drinking and the persistence of pleasure', *Probation Journal*, Vol. 54: 309–326.
Measham F. (2006) 'The new policy mix: Alcohol, harm minimisation and determined drunkenness in contemporary society', *International Journal of Drug Policy*, Special Edition: Harm Reduction and Alcohol Policy, Vol. 17: 258–268.
Measham, F. (2008) 'The Turning tides of intoxication: Young people's drinking in the 2000s', *Health Education*, Vol. 108: 207–222.
Measham, F., Aldridge, A. and Parker, H. (2001) *Dancing on Drugs: Risk, Health and Hedonism in the British Club Scene*. Chippenham: Free Association Books.
Meethan, K. (2001) *Tourism in a Global Society: Place, Culture and Consumption*. London: Palgrave Macmillan.
Miles, S. (2000) *Youth Lifestyles in a Changing World*. Philadelphia: Open University Press.
Miles, S. (2010) *Spaces for Consumption*. London: Sage.
Monaghan, L. (2001) 'Looking good, feeling good: The embodied pleasures of vibrant physicality', *Sociology of Health & Illness*, Vol. 23 (3): 330–356.
Opperman, M. and Chon, K.S. (1997) *Tourism in Developing Countries*. London: Thomson.
O'Neill, M. and Seal, L. (2012) *Transgressive Imaginations: Crime, Deviance and Culture*. London: Palgrave Macmillan.
O'Reilly, K. (2000) *The British on the Costa Del Sol: Transnational Identities and Local Communities*. London: Routledge.
Parker, H., Williams, L. and Aldridge, J. (2002) 'The normalisation of "sensible" recreational drug use: Further evidence from the North West England longitudinal study', *Sociology*, Vol. 36 (4): 941–964.

Patton, Q. (1990) *Qualitative Evaluation and Research Methods*, 2nd ed. Newbury Park, CA: Sage.
Payeras, M., Alcover, A., Alemany, M., Jacob, M., Garcia, A. and Martinez-Ribes, L. (2011) 'The economic impact of charter tourism on the Balearic economy', *Tourism Economics*, Vol. 17 (3): 625–638.
Pearce, P. (2005) *Tourist Behaviour: Theories and Conceptual Schemes*. Clevedon, UK: Channel View.
Presdee, M. (2000) *Cultural Criminology and the Carnival of Crime*. London: Routledge.
Pritchard, A. and Morgan, N.J. (2000) 'Privileging the male gaze: Gendered tourism landscapes', *Annals of Tourism Research*, Vol. 27 (4): 884–905.
Pritchard, A. and Morgan, N.J. (2005) 'On location: Reviewing bodies of fashion and places of desire', *Tourist Studies*, Vol. 5 (3): 283–302.
Raby, R. (2010) 'Public selves, inequality and interruptions: The creation of meaning in focus groups with teens', *International Journal of Qualitative Methods*, Vol. 9 (1): 1–15.
Redhead, S. (1990) *The End of the Century Party: Youth and Pop towards 2000*. Manchester, UK: Manchester University Press.
Redhead, S. (1993) *Rave Off: Politics and Deviance in Contemporary Youth Culture*. Aldershot, UK: Avebury.
Redmon, D. (2003) 'Playful deviance as an urban leisure activity: Secret selves, self-validation, and entertaining performances', *Deviant Behavior*, Vol. 24 (1): 27–51.
Ritchie, J. and Spencer, L. (2004) 'Qualitative data analysis for applied policy research' in A. Bryman and R. Burgess (eds) *Analysing Qualitative Data*. London: Routledge.
Ritzer, G. (2000) *The McDonaldization of Society*, Millenium edition, Thousand Oaks, CA: Pine Forge.
Rojek, C. (2000) *Leisure and Culture*. Basingstoke: Palgrave Macmillan.
Rojek, C. (2005) *Leisure Theory: Principles and Practice*. London: Palgrave Macmillan.
Sanders, T. (2005) *Sex Work*. Cullompton, UK: Willan.
Sanders, T. (2008) *Paying for Pleasure: Men Who Buy Sex*. Cullompton, UK: Willan.
Schor, J. (1992) *The Overworked American*. New York: Basic Books.
Schor, J. and Holt, D.B. (eds) (2000) *The Consumer Society*. New York: New York Press.
Skeggs, B. (2004) *Class, Self, Culture*. London: Routledge.
Skey, M. (2011) *National Belonging and Everyday Life: The Significance of Nationhood in an Uncertain World*. Basingstoke, UK: Palgrave.
Slater, D. (1997) *Consumer Culture and Modernity*. Cambridge: Polity Press.
Smith, O. (2012) 'Easy money: Cultural narcissism and the criminogenic markets of the night-time leisure economy' in S. Winlow and R. Atkinson (eds) *New Directions in Crime and Deviancy*. London: Routledge.
Standing, G. (2011) *The Precariat: The New Dangerous Class*. London: Bloomsbury Academic.
Thomas, M. (2005). 'What happens in Tenerife stays in Tenerife: Understanding women's sexual behaviour on holiday', *Culture, Health and Sexuality*, Vol. 7 (6): 571–584.

Thurnell-Read, T. (2011) 'Off the leash and out of control': Masculinities and embodiment in Eastern European stag tourism', *Sociology*, Vol 45 (6): 977–991.

Thurnell-Read, T. (2012) 'Tourism place and space: British stag tourism in Poland', *Annals of Tourism Research*, Vol. 39 (2): 801–819.

Treadwell, J., Briggs, D., Winlow, S. and Hall, S. (2012) 'Shopocalypse now: Consumer culture and the English riots of 2011', *The British Journal of Criminology*, Vol. 53 (1): 1–17.

Turner, T. and Briggs, D. (2011) 'Understanding the appeal of risk for British youth on holiday in Ibiza: Some ethnographic observations', *European Journal of Youth and Child Research*, Vol. 7: 91–96.

Tutenges, S. (2009) 'Safety problems among heavy-drinking youth at a Bulgarian nightlife resort', *International Journal of Drug Policy*, Vol. 20 (5): 444–446.

Tutenges, S. (2010). *Louder! Wilder! Danish Youth at an International Nightlife Resort*. Copenhagen: Department of Sociology, Copenhagen University.

Tutenges, S. (2012) 'Nightlife tourism: A mixed methods study of young tourists at a nightlife resort', *Tourist Studies*, Vol. 12 (2): 135–155.

Uriely, N. and Belhanssen, Y. (2006) 'Drugs and risk-taking in tourism', *Annals of Tourism Research*, Vol. 33(2): 339–359.

Urry, J. (1990) *The Tourist Gaze*. London: Sage.

Van Maanen, J. (1988) *Tales from the Field: On Writing Ethnography*. Chicago: University of Chicago Press.

Veblen, T. (1994) *The Theory of the Leisure Class*. Harmondsworth, UK: Penguin.

Warr, D.J. (2005) 'It was fun ... but we don't usually talk about these things: Analyzing sociable interaction in focus groups', *Qualitative Inquiry*, Vol. 11: 200–225.

Webb, J., Schirato, T. and Danaher, G. (2002) *Understanding Bourdieu*. London: Sage.

Wilson, B. (2006) *Fight, Flight or Chill: Subcultures, Youth and Rave into the 21st Century*. Montreal: McGill-Queen's University Press.

Winlow, S. (2001) *Badfellas: Crime, Tradition and New Masculinities*. Oxford: Berg.

Winlow, S. and Hall, S. (2006) *Violent Night: Urban Leisure and Contemporary Culture*. Oxford: Berg.

Winlow, S. and Hall, S. (2009) 'Living for the weekend: Youth identities in northeast England', *Ethnography*, Vol. 10 (1): 91–113.

Young, J. (1999) *The Exclusive Society: Social Exclusion, Crime and Difference in Late Modernity*. London: Sage.

Young, J. (2011) *The Criminological Imagination*. Cambridge: Polity Press.

Žižek, S. (2002) *Welcome to the Desert of the Real*. London: Verso.

Žižek, S. (2008) *In Defence of Lost Causes*. London: Verso.

Žižek, S. (2011) *Living in the End Times*. London: Verso.

Index

24/7 party, 91

affordablility in Ibiza, 83, 90
'afterparties', 131, 144
airline companies' offers, 228
airport, holiday hype at, 73–5
alcohol
    consumption, 7, 12, 13, 20–1, 34, 75, 103–4, 116, 132, 163, 208
    deals, 74, 150, 191, 220, 228
    spend much money on, 142
    tempting offers on flights, 74
all-inclusive hotels, 48, 64, 149
anal sex, 222
attraction of Ibiza, 49, 93, 110, 220
Ayia Napa, 3, 11, 64, 81, 89, 109, 117, 118, 155

The Bachelor, 59, 60, 61, 217
Bakhtin, M., 35
Balearic beat, 44, 202–3, 220
Balearic Islands, 11, 12
banal nationalism, 98, 184, 220
bars, 7, 17, 18, 20, 22, 41, 44, 51, 64, 88, 91, 98, 126, 127, 137, 143, 144, 183, 198
Bar Wars, 64
BBC Radio, 1, 203
beach, 7, 9, 41, 46, 61, 83, 89, 116, 136, 198, 222
beach clubs, 49, 89–91, 95, 121, 124, 125, 127, 131, 140, 144, 145, 146–9, 155, 156, 219, 220, 228
beachgirls, 56–7, 83
beach hotels, 129, 213. 220, 228
'below life' reflection, 189, 194–7, 205, 209, 225
Benidorm, 64
Billig, M., 98
'binge drinking', 12, 20, 173, 199
Blackman, S., 62
Blackpool, 79
Blackshaw, T., 34

blatant advertisement, 64, 65, 92, 228
blow job, 9, 103, 139, 161, 170, 222, 235
Blue Marlin, 125, 146, 147–8, 156, 157, 180, 219, 228
boat parties, 66, 131, 144–5, 171
Bodrum (Turkey), 11
booze cruises, 41, 58, 124, 131, 137, 140, 141, 144–5, 158, 173, 199, 201–2, 228
booze-sellers, 168
Bora Bora beach, 71, 88, 113, 114, 155, 156, 180
Bramham, P., 36
branded commodities, 32
Brighton, 79
Britain's Got Talent, 59, 61, 224
British tourists abroad, 1, 9, 11–12, 56, 212, 216
    deviance and risk behaviours, 209–10
    populist media portrayals of, 62–5

Cadena SER, 17
Café del Mar, 42, 181
Café Mambo, 42, 88, 145, 181, 183
Can Misses hospital, 12, 131, 132
Capital FM, 61
capitalismo extremo, 159, 220–3
    awareness about holidays, 170–1
    escaping chavdom, 179–81
    experienced 'Ibiza goers', 181–2
    and ideology of 'living the dream' ideology, 220–3
    losers of, 184–6
    playful deviance and risk, consequences of, 172–7
    pleasure in pain, 177–9
    space, deviance and risk, 160–70
    winners of, 182–4
careerless work, 31, 32
carnivalesque moments, 35
'casual leisure', 32

casual workers, 6, 12–13, 17, 87, 91–5, 117, 124, 133–4, 137, 140, 145, 185, 222
Chaney, D., 33
'chavs', 81, 85, 87, 88, 180, 215
cheating, 104, 161, 192, 193
class dimensions of holiday space, 79
Club 18-30 holiday, 81, 82, 83
  package holiday, 41, 49, 83, 118, 212
  reps, 84
clubbing, 33–4, 62, 64, 86, 87, 161
club crowds, 33, 220
club dancer, 222
Clubland, 166, 191, 218
Club Reps, 63, 64, 216
clubs, 8, 22, 42, 51, 86, 89, 144, 146
cocaine users, 12
commercialisation, 50, 53, 79, 82, 88, 183, 203, 213, 218
commission-only contracts, 151
commodification, 23, 40, 79, 123, 183, 213, 214
  of nostalgia, 189, 202–5, 215
  of status stratification, 154–8, 218–20
condensed partying, 57–62
consent procedures, 19–20
conspicuous consumption, 31, 79, 86, 89, 154, 211, 218
consumer culture, 37, 120
  growth of, 30–3
  structuro-culturo framework of, 37
'consumtariat', 210
Costa del Sol, 62, 64
'crude' behaviours, 87
'cultural experience', 80, 108, 110, 115, 116, 117, 121
cultural mediums, 32, 62
'cultural trips', 140, 219

daily routines and responsibilities, 53–6
data collection methodology, 16
  confidentiality and anonymity, 19–20
  consent procedures, 19–20
  data analysis, 25
  focus groups, 22

limitations, 25–6
  observations, 22–5
  open-ended interviewing, 25
  sample, 18
  sampling methodology and establishing rapport, 18–19
death in Ibiza, 218, 222
decoupling, 29, 98
demand
  of critical assessment, 208
  for drugs, 136
  for sex, 139
deviance and risk behaviours in Ibiza, 209–10
DJs, 49, 61, 141, 148, 153, 156, 167, 200–5, 213, 222
dominant culture, 33
drinking, 3, 8, 17, 20–1, 64–5, 90, 98, 131, 160, 190, 192, 198, 215
  cheap drinks, 57
  drinking stories and creating 'memories', 56–7
  every day on holiday, 172–3
  at home, 55
  on plane, 111
  pre-drinking, 56, 67, 141
  to weekends, 54
drinking strip, 7, 24, 37, 41–2, 151
  Southside Crew on, 160–70
drugs, 133, 136, 138, 161
  and crime, 124
drug-taking, 3, 7, 26, 55, 98, 143, 144, 215
drunken participants, managing relationship with, 21
dualism in the self, 189

'easy birds', 78
easyJet flights, alcohol deals on, 73, 74
economic inequalities, 34–5
economic role of tourism, in Europe, 10
ecstasy users, 12
Eden, 89, 91, 161, 164, 166–7, 171, 179, 219
EHIC (European Health Insurance Card), 132
emic reflection, 107

end goals, 28, 94
enjoyment, 58, 70, 100, 178, 182
Es Paradis, 41, 87, 89, 161, 164, 166, 167, 168, 171, 178, 179, 182, 219
Es Pla, 41
ethnographic research, 2, 20, 208
ethnographic researchers, 19, 21
European Union (EU), tourism across, 9–11
excessive consumption, 3, 14, 37, 55, 65, 78, 92, 99, 113, 116, 118, 120, 129, 131, 145, 155, 205, 212, 216, 218, 221, 225, 228
  observing the practice of, 19
  patterns of, 58
  trajectory of, 81–6
excessive drug and alcohol use, 131–2
'expectations of youth', 104

Facebook, 9, 18, 19, 56, 66–9, 70, 72, 76, 81, 110, 144, 178, 197, 201, 202, 206, 221
Faliraki, 3, 11, 63, 81, 82, 109, 117, 131, 155, 198
'fat birds', 84
Featherstone, M., 35
female groups, 18, 49, 113, 177, 217, 218, 236
Ferrell, J., 28
First Choice, 131
first night, 98, 110, 111–19, 113, 159–60, 169–70, 175, 176, 190, 192, 222, 228
'first timers', 170–1
focus group, 2, 16–19, 22, 26, 175, 192, 198
Foreign and Commonwealth Office (FCO), 130
freedom *versus* unfreedom, 3
Freeview television, 61
'free will' choices, 28
'free will' decisions, 14, 114
'fun' of Ibiza, 6

gendered holiday space, 26
gender relations, 26, 34, 153, 187, 217
'good life', 6, 59, 65, 75, 76, 91, 95, 98, 105, 107, 121, 130, 142, 155, 181, 184, 187, 189, 197, 200, 202, 209, 212, 223, 224
'great moments', recalling, 188
group drinking, 21
group dynamics on holiday, 215–18
Guardia Civil, 135, 240
The Guardian, 62–3
Gümbet (Turkey), 11

habitus, 3, 6, 14, 19, 25, 51, 53, 108, 110, 114, 115, 116, 121, 122, 160, 176, 179, 187, 207, 212, 214, 216, 219
'hacer pingüino', 139
haphazard spending, 172, 221, 227
Harvey, D., 80
Hayward, K., 32, 34
health and wellbeing issues, 10–11
health system, 129–32
Heart (radio station), 61
heavy drinking, 20
Hedkandi events, 204–5
hedonism, 4, 35, 37, 55, 80, 105, 108, 129, 187, 189, 224
hegemonic masculine positions, 217
history of holiday, 79–81
Hobbs, D., 34
holiday, 5, 18, 120, 209–10, 212, 214, 223
  group dynamics on, 215–18
  history of, 79–81
  as a social occasion, 2
'holiday capital', 82, 110, 115, 116, 118, 119, 121, 180, 212
holiday career, 3, 78, 81–6, 87, 95, 212–14
  long road, 82–5
  short cut, 82, 85–6
holiday destinations, 14, 78
'holiday hype', 65–75
  at airport, 73–5
  on Facebook, 66–9
  in shops and gym, 69–73
holiday pinnacle, 82
holiday spaces, 26, 79, 218
home characters, 102
home identities, loosening of, 102, 103

252 *Index*

home lives and holiday hype, 52, 97, 107
  'Brits abroad', populist media portrayals of, 62–5
  daily routines and responsibilities, 53–6
  drinking stories and creating 'memories', 56–7
  'holiday hype', 65–75
  'powering through' the weekend, 57
  seasoned weekenders, 57–62
  socialising intoxication, 56–62
hospitalisations, 131, 222
hotels, 41, 44, 46, 48, 124, 125, 146–9
Hyde, A., 22
hyperconsumption, 98, 99, 107, 108, 111, 113, 115, 121, 152, 155, 158, 160, 170, 172, 174, 189, 190, 193, 200, 205, 211, 216, 217, 221, 224

Ibiza, 1, 4, 49, 52, 124, 171, 213, 227
  British deviant and risk behaviours in, 12–13
  British tourists, 49–50
  choosing, 85
  commercial evolution of San Antonio, 43–9
  concentrated spaces designed for consumption, 140–54
  constructing, 77–96
  destination San Antonio, 40–3
  drugs, police and criminal justice system, 133–6
  ecstasy users in, 12
  'fun' of, 6
  health system, 129–32
  ideological imagery of, 108–10
  ideology behind, 195
  informal economy, 136–40
  infrastructure of, 125–32
  marketisation of, 4, 49, 213, 227
  mass commercialisation of, 218
  rebalancing, 225–7
  short-cutting the holiday career, 85–6
  spanish residents and local tourism firms, 126–9
  uniqueness of, 78

Ibiza 24/7, 129, 130, 131, 227, 228
'Ibiza experience', 90, 155
'Ibiza goers', 87
  experienced, 181–2
Ibiza Rocks, 42, 46, 63, 120, 124, 125, 126, 131, 141, 146, 155, 219
Ibiza Uncovered, 63, 64, 216
identity, 99
  holiday relations, anonymous familiarity of, 102–5
  home identity, loosening of, 102, 103
  internal tug of war on, 223–5
  loosening, 103
  reversal, 189
  self-deconstruction, 105–8
  self-identity, 108
  shift in, 97
  time folding and new permissiveness, 101–2
ideological imagery of Ibiza, 108–10
ideology, 36–8, 103, 109, 195
The Inbetweeners, 65, 216
independent travel, 80
individuality, 99, 119, 121, 171, 172
'inexperienced lads', 85, 223
informal economy, 95, 134, 136–40, 158, 172, 182, 187, 221
inter-group relations, 173
international holiday resorts, young Europeans at, 9–10
international tourism
  demand for, 80
  growth in, 80
intoxication, 19, 216, 221
  culture of, 34–6
  socialising, 56–62
intra-class distinction, 81
invisible status ladders, 219
Izola (Slovenia), 11

'job interview', 92
jouissance, 36, 169, 177, 189, 217, 218, 224

Kavos, 11, 64
Kavos on Corfu (Greece), 11
Kent youth, 198, 200–1
Ketamine, 93, 142, 149, 199

Kevin and Perry Go Large, 65, 216
Kos, 83, 109

'lads' holiday, 85, 200
Laganas on Zante (Greece), 11
lapdancers, 23, 217
last day, approaching,192–4
last night, approaching, 190–2
'last night of madness', 192
leakage, 158
leisure, 211
leisure careers, 31
leisure class, 31, 210–12, 216, 219, 221
leisure pursuits, 30–3
'liberation', 98, 114, 217
'lifestyle', 5, 33, 61
'living the dream' in Ibiza, 6, 61, 182–4, 185–6, 197–202, 213, 220–3
long road, taking (to Ibiza), 82, 213, 215

Made in Chelsea, 59, 60
Magaluf, 78, 81, 97, 100, 230–1
male groups, 18, 151, 177, 217–18
Malia, 3, 11, 64, 65, 83, 84, 85, 89, 118, 155, 172
marketisation of Ibiza, 4, 49, 213, 227
Marmaris (Turkey), 11
masculine identities, 161
  sustaining, 34, 217
mass commercialisation of Ibiza, 203, 218
mass tourism, 43–4, 80, 87
MasterCard, 72, 221
MCAT, 93
McDonaldisation thesis, 36–7
Meethan, K., 23
mono-tourism, 46
'most messy' night, 111
motivation, 79, 212
music, 124, 203, 213
music style conflict, 218
Médico Galeno, 132

narcissistic individualism, 218
neoliberal consumer capitalism, 210, 226
'neo-tribe', 220

new permissiveness, 98, 99, 102, 105, 122, 223
  in identity, 101–2
nightclub, 8, 57, 181
nightlife scenes, 10
night-time economy (NTE), 34
non-stop 24/7 party, 91
'nostalgia', 214
  commodification of, 215

observations, 22–5
Ocean Beach Club Ibiza, 126, 127, 146, 147–8
off-licences, 46, 51
one-to-one, open-ended interviews, 17
'ontological insecurity', 29
'ontological security', 5
open-ended interviewing, 25
over-intoxication, 228

Pacha, 62, 89, 91, 93, 180, 182
package holidays, 11, 80
participant observation study, in holiday resort, 7
participants' *habitus*, 19
partying, 9–10, 57–62, 118
Patton, Q., 21
Pearce, P., 212
penultimate day, 190–2
permissiveness, new, 98, 99, 102, 105, 122, 223
  in identity, 101–2
personal fantasies, 107–8
personal safekeeping, 130
plane, party on, 228
Platja d'en Bossa, 87–9
playful deviance and risk, consequences of, 172–7
pleasure class, 210–12, 215, 216, 219, 221
Policía Local of San Antonio, 134–5
Policía Nacional, 135, 240
political economy, 124
  concentrated spaces designed for consumption, 140–54
  booze cruises, 144–5
  private hotels/beach clubs, 146–9
  Superclubs, 141–4
  West End, 149–54

political economy – *continued*
  informal economy, 136–40
  island's infrastructure, 125
    drugs, police and criminal justice system, 133–6
    health system, 129–32
    spanish residents and local tourism firms, 126–9
  spatial commodification of status stratification, 154–8
populating holiday destinations, 78
postmodern identities, 29–30
postmodern world, 27–9
powering through the weekend, 57
'predrink', 56
'priceless' moments, 100
'Priceless' TV adverts, 72
private hotels, 46, 124, 131, 141, 144, 146–9
private parties, 147
prostitutes, 23, 139–40, 217
prostitution, 133, 158, 221
Public Relations (PRs), 23, 93, 94, 137, 151, 183, 228, 239

'quality birds', 78
Queens Park Ranger, 103

Radio, 1, 61, 203
'random' man, 103
Rascal, Dizzee, 120
rebalancing Ibiza, 225–7
Redhead, S., 33
reflexive double door, 194–7, 224
reflexive evaluation of the self, 189, 200
reflexive methodology, 208–9
reflexive modernisation, 29
'reinventions', 30
reps, 109
resorts, 2, 4, 10, 79, 125, 164, 212, 213
return from holiday, 188
'return-to-Ibiza' decisions, 205–6
risk society, 28
risk-taking, 25, 209–10, 223
Ritzer, G., 360
robberies, 222
Ryanair flights, 73

sameness, superseding, 108
San Antonio, 1, 117, 219, 226
  all-inclusive hotels, 149
  'beachgirls', 56–7
  British oriented, 46
  British tourists at, 49–50
  commercial evolution of, 43–9
  competition for, 48
  crime, disorder and drugs in, 133–6
  destination, 40–3
  drinking stories and creating 'memories', 56–7
  first timers in, 170–1
  in informal economy, 138
  Kent youth in, 198–9
  mapping of the West End 'drinking strip' in, 24
  mishaps in, 129
  observations in, 22–4
  pilot focus groups and initial ethnographic fieldwork in, 18
  sex in, 170
  Southside Crew in, 7–9, 16
  tourism in, 83, 85, 87–9
  tourism landscape across, 126
  West End 'drinking strip' in, 24
  women in, 217
San Antonio Bay, 23, 41, 44
seaside resorts, 79
seasoned weekenders, 57–62
'seizing the moment'
  attitude of, 31, 221
  ontology of, 160, 176
self
  general evaluation of, 196
  reflexive evaluation of, 189
self-deconstruction, 105–8, 189, 194, 209, 224
self-identity, 108
self-indulgence, 221
self-injury, 222
self-love, 29
self-rationalisation, 95, 99
self-realisation, 31, 36, 216
semi-naked women, 104
sex, 1, 5, 6, 7, 13, 22, 60, 63, 65, 81, 139, 161, 170, 171, 187, 192, 212, 217
sex industry, 10, 158

Index

shagging women, 166, 170
short breaks, 80
short cutters, 213, 215
short-cutting the holiday career, 82, 85–6
Sky TV, 61
social context, players in, 12
social credibility, 91, 119
social distinction, 4, 34
social envy, 86, 91
social expectations, 55, 103
social gradings, 86
social relations, 105, 223
social scenery, players in, 17
Soul City, 166
Southside Crew, 7–9, 54, 70, 74, 78, 97, 159, 208, 232–8
  on drinking strip, 160–70
  last day, 193
Space (club), 173, 180, 181
space, deviance and risk, 160–70
spanish residents and local tourism firms, 126–32
spatial commodification of status stratification, 154–8, 219
Standing, G., 210
'status inadequacy', 219
status-placing activity, 31
status stratification, 86
  spatial commodification of, 154–8, 218–20, 219
statutory holiday pay, 79
'stay safe' campaign, 130
strip clubs, 20, 151, 152, 222
strippers, 23, 217
structured holiday time, 79
subculture, 33, 220
summer tourist market, 45
The Sun, 63
Sun, Sex and Suspicious Parents, 64, 216
Sunny Beach (Bulgaria), 11, 116
Superclubs, 8, 89, 121, 124, 125, 141–4, 155, 183, 213, 219, 220
Superclub shops, 40, 46
Surreal Brummies, 172–4

Take Me Out, 59, 60–1, 217
Televisión Ibiza Formentera, 17

thematic/ theoretical discussions
  capitalismo extremo, 220–3
  group dynamics on holiday, 215–18
  holiday career, 212–14
  identity, internal tug of war on, 223–5
  pleisure class, 210–12
  rebalancing Ibiza, 225–7
  reducing the risks of tourists, 227–9
  status stratification, spatial commodification of, 218–20
  'youth', concept of, 214–15
theoretical framework, 27
  clubbing, 33–4
  consumer culture and leisure pursuits, growth of, 30–3
  deviance and risk-taking, 33–6
  ideology, 36–8
  intoxication, culture of, 34–6
  postmodern identities, 29–30
  postmodern world, 27–9
  work-based identities, decline of, 30–3
Thomas Cook (tourist company), 82, 131
time folding, 100–2, 204, 214–15
'time out', 10, 189–90
tourism across the European Union (EU), 9–11
'tourist', 80
  and casual worker, 91–5
  reducing the risks of, 227–9
tourist spaces, 218
tourist spending, 222
tourists' behaviour, tackling, 133
TOWIE, 59, 60
transgression, 34, 35, 36, 53–6, 215
transportation, 80
'travel career', 212
'traveller', 80
'travelling', 116, 117, 214
'true freedom', 114

UK seaside resorts, 79
unfreedom, 114, 115, 119, 211
  first night, 111–19
  transgressing, 119–21
'unpredictables', 57, 85–6, 99
unprotected sex, 10, 12, 13, 218

urban working classes, 79
Ushuaia, 90–1, 131, 141, 146, 155

Veblen, T, 31–2
The Villa, 64, 216
violence, 19, 34, 183, 216, 222
VIP areas, 127, 141, 156, 157

Wagg, S., 36
Webb, J., 3
weekend nights, 62, 120
'weekend self', 224
weekend 'person', 105, 108
West End, 41, 42, 124, 133, 149–54, 166–9, 183, 205

'drinking strip', 7, 8, 23, 24
  prostitutes in, 139–40
  women at, 139–40, 151, 153–4, 170
work-based identities, decline of, 30–3
working-class groups, 34, 81
WTF Boat Party, 144

X Factor (programme), 59, 60, 61, 92

'youth', concept of, 212–13, 214–15

Zante, 81, 83
Žižek, S., 14, 37, 210

Printed by Printforce, the Netherlands